HOLOMORPHIC AUTOMORPHISM GROUPS IN BANACH SPACES: AN ELEMENTARY INTRODUCTION

NORTH-HOLLAND MATHEMATICS STUDIES 105
Notas de Matemática (97)

Editor: Leopoldo Nachbin

Centro Brasileiro de Pesquisas Fisicas,
Rio de Janeiro,
and University of Rochester

NORTH-HOLLAND – AMSTERDAM • NEW YORK • OXFORD

HOLOMORPHIC AUTOMORPHISM GROUPS IN BANACH SPACES: AN ELEMENTARY INTRODUCTION

José M. ISIDRO

Facultad de Matemáticas
Universidad de Santiago de Compostela
Spain

and

László L. STACHÓ

Bolyai Intézet
Szeged
Hungary

1985

NORTH-HOLLAND – AMSTERDAM • NEW YORK • OXFORD

ISBN: 0 444 87657 X

Publishers:
ELSEVIER SCIENCE PUBLISHERS B.V.
P.O. BOX 1991
1000 BZ AMSTERDAM
THE NETHERLANDS

Sole distributors for the U.S.A. and Canada:
ELSEVIER SCIENCE PUBLISHING COMPANY, INC.
52 VANDERBILT AVENUE
NEW YORK, N.Y. 10017
U.S.A.

Library of Congress Cataloging in Publication Data

Isidro, José M.
Holomorphic automorphism groups in Banach spaces.

(North-Holland mathematics studies ; 105) (Notas de matemática ; 97)
Bibliography: p.
1. Holomorphic functions. 2. Automorphisms.
3. Banach spaces. I. Stachó, László L. II. Title.
III. Series. IV. Series: Notas de matemática (Amsterdam, Netherlands) ; 97.
QA1.N86 no. 97 [QA331] 510 s [515.9'8] 84-21164
ISBN 0-444-87657-X (U.S.)

PRINTED IN THE NETHERLANDS

PREFACE

Since the early 70's, there has been intensive development in the theory of functions of an infinite number of complex variables. This has led to the establishment of completely new principles (e.g. concerning the behaviour of fixed points) and has thrown new light on some classical finite dimensional results such as the maximum principle, the Schwarz lemma and so on. Perhaps the most spectacular advances occurred in connection with the old problem of the determination of the holomorphic automorphisms of complex manifolds.

This book is based on the introductory lectures on this latter field delivered at the University of Santiago de Compostela in October 1981 by the authors. Originally, it was planned as a comprehensive postgraduate course relying on a deep knowledge of holomorphy in topological vector spaces and infinite dimensional Lie groups. However, seeing that some of the undergraduate students were mainly interested in the study of bounded domains in Banach spaces, the authors restricted their attention to these aspects. This proved to be a fortunate idea. We realized that by combining the methods of the theories developed independently by W. Kaup and J.P. Vigué with minor modifications, even the main theorems could be derived. This was achieved in a self-contained way from the most fundamental principles of Banach spaces (such as the open mapping theorem), elementary function theory and the pure knowledge of the Taylor series representation of holomorphic maps in this setting. It may often happen in teaching mathematics that avoiding the introduction of strong tools leads to abandoning natural heuristics. Probably, this is not the case now. It is enough to recall how deeply the early development of the theory of finite dimensional Lie groups and Lie algebras was inspired in Cartan's investigation of the structure of symmetric domains. Moreover, we think that this approach to the automorphism groups of Banach space domains may also serve as motivating and illustrative material in introducing students to the theory of Lie groups and complex manifolds.

The text is divided into eleven chapters. In chapter 0 we establish the terminology, and some typical examples of later importance (e.g. the Möbius group) are studied. In chapter 1 we show the main topological consequences of the Cauchy estimates of Taylor coefficients for uniformly bounded families of holomorphic mappings. These considerations are continued in chapter 2 and applied specifically to the case of the automorphism group, concluding with the topological version of Cartan's uniqueness theorem. The global topological investigations finish in chpater 3, where the Carathéodory distance is introduced to obtain the completness properties of the group AutD. In chapter 4 a completely elementary introduction to Lie theory begins by showing where one-parameter subgroups come from. Chapter 5 is devoted to a description of the Banach Lie algebra structure of complete holomorphic vector fields in order to lay the foundation of chpater 6, in which the Banach Lie groups structure of AutD is studied. In chpaters 7 and 8 we discuss the basic theory of circular domains and determine explicitly the holomorphic automorphism group of the unit ball of several classical Banach spaces. In chapter 9 we introduce the reader to another fruitfully developing branch of these researches by proving Vigué's theorem on the Harish-Chandra realization of bounded symmetric domains. Finally, in chapter 10 and elementary introduction of the Jordan approach to bounded symmetric domains is presented and the convexity of the Harish-Chandra realization is proved.

We would like to express our sincere acknowledgement to Prof. L. Nachbin who suggested the idea of writing these notes and who, together with Prof. E. Vesentini, introduced the authors to infinite dimensional holomorphy and this fascinating branch of mathematics.

Thanks are also due to M. Teresa Iglesias for the careful typing.

The authors, August 1984.

J.M. Isidro
Santiago de Compostela
Spain.

L.L. Stachó
Szeged
Hungary.

TABLE OF CONTENTS

CHAPTER 0

PRELIMINARIES

Throughout what follows, E and E_1 denote complex Banach spaces whose norms will be represented indistinctly by $\|\cdot\|$, and D is a bounded domain in E.

<u>0.1. DEFINITION</u>. *A mapping* $f: D \to E_1$ *is said to be holomorphic if, for every* $a \in D$, *we have*

$$f(a+h) = \sum_{n=0}^{\infty} f_a^{(n}(h,\dots,h)$$

in a neighbourhood of a.

Here, for every $n \in \mathbb{N}$,

$$(0.1) \qquad f_a^{(n}: (h_1,\dots,h_n) \to \frac{1}{n!} \frac{\partial^n}{\partial t_1 \dots \partial t_n}\Big|_0 f(a+t_1h_1+\dots+t_nh_n)$$

is a continuous n-linear operator from E^n into E_1. Remark that, for $n \in \mathbb{N}$ and $h \in E$, we have

$$(0.2) \qquad f_a^{(1}(h,\dots,h) = \frac{1}{n!} \frac{\partial^n}{\partial t^n}\Big|_0 f(a+th)$$

The family of all holomorphic mappings from $D \subset E$ into a set $D_1 \subset E_1$ is denoted by $\mathrm{Hol}(D,D_1)$. When $E = E_1$ and $D = D_1$ we write $\mathrm{Hol}(D)$ instead of $\mathrm{Hol}(D,D_1)$.

<u>0.2. DEFINITION</u>. *A subset* $B \subset D$ *is said to be completely interior to* D, *and we write* $B \subset\subset D$, *if* $\mathrm{dist}(B, \partial D) > 0$.

For $f \in \mathrm{Hol}(D,D_1)$ and $B \subset\subset D$ we define $\|f\|_B$ by means of

$$\|f\|_B =: \sup_{x \in B} \|f(x)\|$$

<u>0.3. DEFINITION</u>. *A net* $(f_j)_{j \in J}$ *in* $\mathrm{Hol}(D,D_1)$ *is said to converge locally uniformly to a mapping* $f \in \mathrm{Hol}(D,E_1)$ *if, for*

every ball $B \subset\subset D$, *we have*

$$\lim_{j \in J} \| f_j - f \|_B = 0$$

We denote by $\mathcal{T}$ the topology on $\mathrm{Hol}(D,D_1)$ of local uniform convergence over D. If a net $(f_j)_{j \in J}$ in $\mathrm{Hol}(D,D_1)$ is locally uniformly convergent to $f \in \mathrm{Hol}(D,E_1)$, we write

$$\mathcal{T}\text{-}\lim_{j \in J} f_j = f$$

0.4. EXERCISE. (a) Let E be the Banach space ℓ^1 and set $f(\zeta_1,\zeta_2,\ldots,) =: \sum_{n=0}^{\infty} n\zeta_n^n$ for $\zeta = (\zeta_1,\zeta_2\ldots) \in \ell^1$. Show that the function $f\colon E \to \mathbb{C}$ is holomorphic on the whole space E and that f is not bounded on the open unit ball B(E) of E. Thus we may have $\| f \|_B = \infty$ even if $f \in \mathrm{Hol}(D,E_1)$ and $B \subset\subset D$.

(b) Is $\mathcal{T}$ a metrizable topology?.

0.5. DEFINITION. *A mapping* $f \in \mathrm{Hol}(D)$ *is said to be an automorphism of* D *if there exists* $g \in \mathrm{Hol}(D)$ *such that*

$$fg = \mathrm{id}_D = gf$$

Here fg stands for the composite of the mappings f and g, and id_D represents the identity mapping of D. The family of all automorphisms of D is represented by AutD.

0.6. EXERCISE. (a) Prove that a mapping $f\colon D \to E$ satisfies $f \in \mathrm{Aut}D$ if, and only if, f is a surjective bijection of D and, for every $a \in D$, the operator $f_a^{(1}$ is invertible.

(b) Can the assumption concerning the invertibility of $f_a^{(1}$ be weakened?.

(c) Show that AutD with the usual law of composition is a group.

0.7. EXAMPLE. Let Δ be the open unit disc of $\mathbb{C}$ and, for $k,u \in \mathbb{C}$ with $|k| = 1$ and $|u| < 1$, let us define $M_{k,u}$ as the restriction to Δ of the Möbius transformation

$$M_{k,u}: \zeta \to k\,\frac{\zeta+u}{1+\bar{u}\zeta}, \qquad \zeta\in\bar{\mathbb{C}}$$

Then, the following result holds:

0.8. THEOREM. *The group* AutΔ *is given by*

$$\mathrm{Aut}\Delta = \{M_{k,u};\ k,u\in\mathbb{C}\quad |k|=1\quad |u|<1\}.$$

Proof: First, let us observe that we have

$$K_{k,u} = M_{k,0}\,M_{1,u}\qquad M_{1,u}\,M_{k',1} = M_{k',u\bar{k}'}\qquad M_{1,u}\,M_{1,u'} = M_{v,w}$$

where $v=:\dfrac{1+u\bar{u}'}{1+\bar{u}u'}$ and $w=:\dfrac{u+u'}{1+u\bar{u}'}$ satisfy $|v|=1$ and $|w|<1$. Moreover,

$$M_{1,u}\,M_{1,-u} = id_\Delta = M_{1,-u}\,M_{1,u}\quad\text{and}\quad M_{k,0}\,M_{\bar{k},0} = id_\Delta = M_{\bar{k},0}\,M_{k,0}$$

Thus, the Möbius transformations form a group with regard to composition and every $M_{k,u}$ is biholomorphic on $\bar{\mathbb{C}}$. Therefore we have $M_{k,u}(\partial\Delta) = \partial M_{k,u}(\Delta)$. Let $\xi\in\partial\Delta$ and $M_{k,u}$ be arbitrarily fixed; since $|k| = |\zeta| = 1$, we have

$$|M_{k,u}(\zeta)| = \left|\frac{\zeta+u}{1+\bar{u}\zeta}\right| = \left|\frac{\bar{\zeta}(\zeta+u)}{1+u\bar{\zeta}}\right| = \left|\frac{1+u\bar{\zeta}}{1+\bar{u}\zeta}\right| = 1$$

that is, $M_{k,u}(\partial\Delta)\subset\partial\Delta$. Thus $\partial M_{k,u}(\Delta)\subset\partial\Delta$. On the other hand we have $M_{k,u}(\zeta)\neq\infty$ for all $\zeta\in\bar{\Delta}$, whence we see that $M_{k,u}(\Delta)$ is a bounded open set in $\mathbb{C}$ whose boundary is contained in $\partial\Delta$. However, one easily checks that the only set with these properties is Δ, i.e., $M_{k,u}(\Delta) = \Delta$ and $M_{k,u}\in\mathrm{Aut}\Delta$.

Conversely, let $f\in\mathrm{Aut}\Delta$ be given and write $u=: f(0)$, $g=: M_{1,-u}f$. Then $g(0) = 0$ so that the Schwarz lemma may be applied to g and g^{-1}. Then, for any $\zeta\in\Delta\setminus\{0\}$, we have $|g(\zeta)|\leq|\zeta|$ and $|\zeta| = |g^{-1}g(\zeta)|\leq|g(\zeta)|$, that is, $|g(\zeta)| = |\zeta|$. Therefore g is linear, so that $g = M_{k,0}$ for some $k\in\partial\Delta$. But then $f = M_{1,u}g = M_{1,u}\,M_{k,0}$ is a Möbius transformation. #

0.9. DEFINITION. *A domain* D *is said to be homogeneous if, for every pair* $a,b\in D$, *there is an automorphism* $f\in \mathrm{Aut}D$ *such that* $f(a)=b$. *If* D *is homogeneous, then we say that* AutD *acts transitively on* D.

It follows that Δ is a homogeneous domain. By Riemann's famous theorem, every proper domain D in $\mathbb{C}$ is *biholomorphically equivalent* to Δ, i.e., there are $f\in \mathrm{Hol}(\Delta,D)$ and $g\in \mathrm{Hol}(D,\Delta)$ such that

$$f(\Delta)=D, \quad g(D)=\Delta, \quad fg=\mathrm{id}_D, \quad gf=\mathrm{id}_\Delta$$

Hence, for every proper domain $D\subset\mathbb{C}$, we can describe explicitely the group AutD by the formula $\mathrm{Aut}D=\{fMf^{-1};\ M\in \mathrm{Aut}\Delta\}$. Let us recall that the mapping f can be expressed in terms of the Green function of the domain D. In particular, each proper domain of $\mathbb{C}$ is homogeneous.

0.10. EXAMPLE. Let E be any complex Banach space and put $D=:B(E)$ for the open unit ball of E. If f is any surjective linear isometry of E, then we have $f|_D\in \mathrm{Aut}D$.

CHAPTER 1

UNIFORMLY BOUNDED FAMILIES OF HOLOMORPHIC MAPS AND LOCALLY UNIFORM CONVERGENCE

§1.- Cauchy majorizations.

Let E and E_1 be complex Banach spaces, $D \subset E$ and $D_1 \subset E_1$ be bounded domains and $a \in D$ be a given point in D. Write $\delta =: \mathrm{dist}(a, \partial D)$. Then,

1.1. PROPOSITION. *(Cauchy estimates). We have*

(1.1) $$\| f_a^{(n} \| \leqslant \left(\frac{e}{\delta} \right)^n \| f \|_D$$

for all $f \in \mathrm{Hol}(D, D_1)$ *and* $n \in \mathbb{N}$.

Proof: By Cauchy's classical integral representation theorem, we have

$$f_a^{(n}(h_1, \ldots, h_n) =$$

$$= \frac{1}{n!} \left(\frac{1}{2\pi i} \right)^n \int_{|\zeta_1| = \ldots = |\zeta_n| = 1} \cdots \int \frac{f(a + \zeta_1 h_1 + \ldots + \zeta_n h_n)}{\zeta_1^2 \ldots \zeta_n^2} d\zeta_1 \ldots d\zeta_n$$

whenever $a + \Delta h_1 + \ldots + \Delta h_n \subset D$. Therefore

$$\| f_a^{(n}(h_1, \ldots, h_n) \| \leqslant \frac{1}{n!} \left(\frac{1}{2\pi} \right)^n (2\pi)^n \| f \|_D$$

for all $h_1, \ldots, h_n \in E$ with $\|h_1\|, \ldots, \|h_n\| < \delta/n$. By the continuity of $f_a^{(n}$, this estimate also holds for $\|h_1\| = \ldots = \|h_n\| = \delta/n$. By the n-linearity of $f_a^{(n}$,

$$\|f_a^{(n}\| = \sup\left\{ \frac{\|f_a^{(n}(h_1, \ldots, h_n)\|}{(\delta/n)^n} ; \ \|h_j\| = \delta/n, \quad j = 1, \ldots, n \right\} \leqslant$$

$$\leq \frac{1}{n!}\left(\frac{n}{\delta}\right)^n \|f\|_D .$$

From the Stirling formula we derive $\frac{n^n}{n!} \leq e^n$, whence the result follows.

#

1.2. EXERCISE. Show by examples in $\mathbb{C}^n$ that, in general, the constant $\frac{n^n}{n!}$ in (1.1) cannot be improved.

1.3. REMARK. *For the "symmetric" terms* $f_a^{(n}(h,\ldots,h)$ *we have the sharper estimate*

$$(1.3) \qquad \|f_a^{(n}\| \leq \left(\frac{1}{\delta}\right)^n \|f\|_D$$

Proof: From (0.2) it follows that

$$f_a^{(n}(h,\ldots,h) = \frac{1}{2\pi i}\int_{|\zeta|=1} \frac{f)a+\zeta h)}{\zeta^{n+1}}\, d\zeta$$

whenever $a+\Delta h \subset D$. Then, as in the proof of (1.1) we obtain

$$\|f_a^{(n}\| = \sup_{\|h\|=1} \|f_a^{(n}(h,\ldots,h)\| \leq \frac{1}{\delta^n}\|f\|_D .$$

#

From the Cauchy estimates we can derive the locally Lipschitzian behaviour of the derivatives of f given by the following:

1.4. PROPOSITION. *Let* a, $b \in D$ *be given and assume that the segment* $[a,b] =: \{a+\lambda(b-a);\ \lambda \in [0,1]\}$ *lies in* D *with* $\mathrm{dist}([a,b], \partial D) = \rho > 0$. *Then we have*

$$(1.3) \qquad \|f_a^{(n}-f_a^{(n}\| \leq (n+1)\left(\frac{e}{\rho}\right)^{n+1} \|f\|_D \|b-a\|$$

for all $f \in \mathrm{Hol}(D,D_1)$ *and* $n \in \mathbb{N}$.

Proof: Set $h_{n+1} =: b-a$. By the Newton-Leibnitz formula we have

$$(f_b^{(n}-f_a^{(n})(h_1,\ldots,h_n) = \int_0^1 \left[\frac{d}{dt} f_{a+th_{n+1}}^{(n}(h_1,\ldots,h_n)\right]dt =$$

$$= \int_0^1 \left|\frac{\partial}{\partial t_{n+1}} \left[{}_0 f^{(n}_{a+th_{n+1}+t_{n+1}h_{n+1}}(h_1,\ldots,h_n)\right] dt = \right.$$

$$= \int_0^1 \frac{\partial}{\partial t_{n+1}} \Big|_0 \frac{\partial^n}{\partial t_1,\ldots,\partial t_n} \Big|_0 f(a+th_{n+1}+t_1h_1+\ldots$$

$$\ldots+t_{n+1}h_{n+1})\,dt$$

which by (0.1) is equal to

$$\int_0^1 (n+1) f^{(n+1}_{a+th_{n+1}}(h_1,\ldots,h_{n+1})\,dt \;.$$

Then, using (1.1) we derive

$$\| f_b^{(n} - f_a^{(n} \| \leqslant \sup_{\|h_1\|=\ldots=\|h_n\|=1} \int_0^1 (n+1) \| f^{(n+1}_{a+th_{n+1}}(h_1,\ldots,h_{n+1} \| dt \leqslant$$

$$\leqslant (n+1) \left(\frac{e}{\rho}\right)^{n+1} \| f \|_D \, \| h_{n+1} \|$$

whence the result follows.

<u>1.5. THEOREM</u>. *Let* $(f_j)_{j\in J}$ *and* f *respectively be a net and an element in* Hold(D,D_1) *and denote by* $B \subset\subset D$ *any ball centered at* $a\in D$. *Then, the following statements are equivalent:*

(a) The net $(f_j)_{j\in J}$ *is uniformly convergent to* f *on* B.

(b) For all $k\in\mathbb{N}$, *we have* $\lim_{j\in J} \| f^{(k}_{j,a} - f^{(k}_a \| = 0$.

Proof: The implication a => b is obvius.

Let r be the radius of B so that $0<r<\delta =: \mathrm{dist}(a,\partial D)$. Given any $n\in\mathbb{N}$, we have for all $h\in E$ with $\|h\| \leqslant r$

$$\| f_j(a+h) - f(a+h) \| = \left\| \sum_{k=0}^{\infty} (f^{(k}_{j,a} - f^{(k}_a)(h,\ldots,h) \right\| \leqslant$$

$$\leqslant \sum_{k=0}^{\infty} \| f^{(k}_{j,a} - f^{(k}_a \| \, \| h \|^k$$

which, by (1.2) is dominated by

$$\sum_{k=0}^{n} \| f^{(k}_{j,a} - f^{(k}_a \| \, r^k + \mathrm{diam}(D_1) \sum_{k=n+1}^{\infty} \left(\frac{r}{\delta}\right)^k$$

Now, given any $\varepsilon>0$ there exists $n_0\in\mathbb{N}$ such that

$$\operatorname{diam}(D_1)\sum_{k=n_0+1}^{\infty}\left(\frac{r}{\delta}\right)^k<\varepsilon/2$$

and, by hypothesis there is $j_0\in J$ such that

$$\sum_{k=0}^{n_0}\|f_{j,a}^{(k}-f_a^{(k}\|\,r^k<\varepsilon/2$$

for all $j\in J$, $j\geq j_0$. This completes the proof.

#

1.6. THEOREM. *Let the balls* B_1 *and* B_2 *be completely interior to* D. *If* $(f_j)_{j\in J}$ *is a net in* $\operatorname{Hold}(D,D_1)$, *the following assertions are equivalent:*

(a) $f_j\to f$ *relative to* $\|\cdot\|_{B_1}$.

(b) $f_j\to f$ *relative to* $\|\cdot\|_{B_2}$.

Proof: Since D is connected, we can find balls $B_0',B_1',\ldots,B_{n+1}'$ such that:

$B_0'\equiv B_1$ and $B_{n+1}'\equiv B_2$

$B_k'\subset\subset D$ for all $k=1,2,\ldots,n$

the center a_k' of B_k' satisfies $a_k'\in B_{k-1}'$ for $k=1,\ldots,n+1$.

As $f_j\to f$ relative to $\|\cdot\|_{B_0'}$, aplying theorem 1.5 to the point a_1' and the ball B_0' instead of and D, we get

$$f_{j,a_1'}^{(r}\to f_{a_1'}^{(r}\quad\text{for all } r\in\mathbb{N},$$

so that $f_j\to f$ relative to $\|\cdot\|_{B_1'}$. After serveral reiterations of the argument we get $f_j\to f$ relative to $\|\cdot\|_{B_{n+1}'}$. Thus

$$f_j\to f \text{ relative to } \|\cdot\|_{B_1} \Rightarrow f_j\to f \text{ relative to } \|\cdot\|_{B_2}.$$

As the roles of B_1 and B_2 may be changed, the proof is complete.

#

1.7. COROLLARY. *The topology* T *on* $\mathrm{Hol}(D,D_1)$ *is metrizable. For any ball* $B\subset\subset D$, $\|\cdot\|_B$ *is a metric on* $\mathrm{Hol}(D,D_1)$ *whose associated topology is* T. *We have* $T\lim f_j = f$ *in* $\mathrm{Hol}(D,D_1)$ *if, and only if, there exists* $a\in D$ *such that* $f_{j,a}^{(k}\to f_a^{(k}$ *for all* $k\in\mathbb{N}$, *or if and only if there exists a ball* $B\subset\subset D$ *such that* $\|f_j - f\|_B \to 0$.

§2.- Continuity of the composition operation.

Let D, D_1 and D_2 be bounded domains in the Banach spaces E, E_1 and E_2. As a first application of the previous theorem we show that the composition of mappings

$$\mathrm{Hol}(D,D_1)\times\mathrm{Hol}(D_1,D_2)\to\mathrm{Hol}(D,D_2)$$

$$(f,g)\to gf$$

is continuous with regard to the topology of local uniform convergence. The way we shall follow is perhaps not the shortest possible but it provides information that turns out to be useful later.

1.8. PROPOSITION. *Let* $f\in\mathrm{Hol}(D,D_1)$ *and* $g\in\mathrm{Hol}(D_1,D_2)$ *be holomorphic mappings whose respective Taylor's series at* $a\in D$ *and* $b=: f(a)\in D$ *are*

$$f(a+h) = f(a) + \sum_{n=1}^{\infty} f_a^{(n}(h,\ldots,h)$$

and

$$g(b+h) = g(b) + \sum_{m=1}^{\infty} g_b^{(m}(h,\ldots,h)$$

Then we have $gf\in\mathrm{Hol}(D,D_2)$ *and*

$$(1.4)\qquad (gf)_a^{(k} = \sum_{m=1}^{k}\ \sum_{\nu_1+\ldots\nu_m=k} g_{f(a)}^{(m}\left[f_a^{(\nu_1},\ldots,f_a^{(\nu_m}\right]$$

for all $k\in\mathbb{N}$.

Here, the detailed interpretation of (1.4)

$$(gf)_a^{(k}: (h_1,\ldots,h_k)\to$$

$$\to \sum_{m=1}^{k} \sum_{\substack{\nu_1+\dots+\nu_m=k \\ \nu_1+\dots+\nu_m\geq 1}} g^{(n}_{f(a)}\left[f_a^{(\nu_1}(h_1,\dots,h_{\nu_1}),\dots,f_a^{(\nu_m}(h_{\nu_{m+1}+1}\dots h_{\nu_m})\right]$$

Proof: We have the following formal expansion for gf about the point $a\in D$

$$(1.5) \qquad gf(a+h) = g(b) + \sum_{m=1}^{\infty} g_b^{(m}[f(a+h)-b;\dots;f(a+h)-b] =$$

$$= g(b) + \sum_{m=1}^{\infty} g_b^{(m}\Big[\sum_{\nu_1=1}^{\infty} f_a^{(\nu_1}(h,..,h);..;\sum_{\nu_m=1}^{\infty} f_a^{(\nu_m}(h,..,h)\Big] =$$

$$= g(b) + \sum_{m=1}^{\infty} \sum_{\nu_1,..,\nu_m=1} g_b^{(m}[f_a^{(\nu_1}(h,..,h);..;f_a^{(\nu_m}(h,..,h)]$$

We point out that (1.5) is uniformly convergent in a suitable neighbourhood of a. Indeed, by (1.1) we have the majorizations

$$\| g_b^{(m}[f_a^{(\nu_1}(h,..,h);..;f_a^{(\nu_m}(h,..,h)] \| \leq$$

$$\leq \| g_b^{(m}\| \, \| f_a^{(\nu_1}\| \cdots \| f_a^{(\nu_m}\| \, \|h\|^{\nu_1+..+\nu_m}$$

$$\leq \left(\frac{e}{\varepsilon}\right)^m \|g\|_{D_1} \left[\left(\frac{e}{\delta}\right)\|f\|_D \|h\|\right]^{\nu_1+..+\nu_m}$$

where $\delta =: \mathrm{dist}(a,\partial D)$ and $\varepsilon =: \mathrm{dist}(b,\partial D_1)$. Hence

$$\infty > \sum_{m=1}^{\infty} \left(\frac{e}{\varepsilon} \|g\|_{D_1}\right)^m \Big[\sum_{n=1}^{\infty}\left(\frac{e}{\delta} \|f\|_D \|h\|^n\right]^m =$$

$$= \sum_{m=1}^{\infty}\left(\frac{e}{\varepsilon}\|g\|_{D_1}\right)^m\Big[\sum_{\nu_1=1}^{\infty}\left(\frac{e}{\delta}\|f\|_D\|h\|\right)^{\nu_1}\Big]\dots$$

$$\dots\Big[\sum_{\nu_m=1}^{\infty}\left(\frac{e}{\delta}\|f\|_D\|h\|\right)^{\nu_m}\Big] =$$

$$= \sum_{m=1}^{\infty}\sum_{\nu_1,..,\nu_m=1}^{\infty}\left(\frac{e}{\varepsilon}\|g\|_{D_1}\right)^m\left(\frac{e}{\delta}\|f\|\,\|h\|\right)^{\nu_1+..+\nu_m} =$$

$$= \sum_{k=1}^{\infty}\sum_{m=1}^{k}\Big(\sum_{\nu_1+\dots+\nu_m=k}^{\infty} 1\Big)\left(\frac{e}{\varepsilon}\|g\|_{D_1}\right)^m\left(\frac{e}{\delta}\|f\|_D\|h\|\right)^k \geq$$

$$\geqslant \sum_{m=1}^{\infty} \sum_{\nu_1,..,\nu_m=1} \| g_b^{(m}[f_a^{(\nu_1}(h,..,h),..,f_a^{(\nu_m}(h,..,h)] \|$$

whenever

$$\sum_{n=1}^{\infty} \left(\frac{e}{\delta} \|f\|_D \|h\| \right)^n < \left(\frac{e}{\varepsilon} \|g\|_{D_1} \right)^{-1}$$

which is satisfied for sufficiently small values of $\|h\|$, say for $\|h\|<\delta_0$. Therefore (1.5) makes sense for all $\|h\|<\delta_0$; moreover, we may write

$$gf(a+h) = g(b) +$$

$$+ \sum_{m=1}^{\infty} \sum_{k=1}^{\infty} \sum_{\substack{\nu_1+..+\nu_m=k \\ \nu_1,..,\nu_m \geqslant 1}} g_b^{(m}[f_a^{(\nu_1}(h,..,h);..;f_a^{(\nu_a}(h,..,h)] =$$

$$= g(b) + \sum_{k=1}^{\infty} \sum_{m=1}^{k} \sum_{\nu_1+..+\nu_m=k} g_b^{(m}[f_1^{(\nu}(h,..,h);..;f_a^{(\nu_m}(h,..,h)]$$

Observe that, for each $k\in\mathbb{N}$, the k-th term of this series is a continuous k-homogeneous polynomial of the variable h. By the uniqueness of the Taylor expansion, this completes the proof.

#

If $E_1,E_2,..,E_n$ and E are complex Banach spaces, we write $\mathcal{L}(E_1,..,E_n|E)$ for the space of all continuous n-linear mappings $L: E_1\times E_2\times..\times E_n\to E$.

<u>1.9. LEMMA.</u> *Let* $(L_j)_{j\in J}$ *be a net in* $\mathcal{L}(E_1,..,E_n|E)$ *with* $L_j\to L$ *and assume that, for* r= 1,2,..,n, $(K_j^r)_{j\in J}$ *are nets in* $\mathcal{L}(E_1^r,..,E_{\nu_r}^r|E_r)$ *with* $K_j^r\to K^r$. *Then we have*

$$L_j(K_j^1,..,K_j^n)\to L(K^1,..,K^n)$$

in

$$\mathcal{L}(E_1^1,..,E_{\nu_1}^1;..;E_{\nu_n}^1,..,E_{\nu_n}^n|E).$$

Proof: Let us write $K^r_{j,0} =: K^r$, $K^r_{j,1} =: K^r_j - K^r$ and similarly $L_{j,0} =: L$, $L_{j,1} =: L_j - L$. We may assume that $\| K^r_{j,s} \|$, $\| L_{j,s} \| \leq \mu$ for all indexes j,r,s. Then

$$L_j(K^1_j,..,K^m_j) = \sum_{\sigma,\sigma_1,..,\sigma_m=0,1} L_{j,\sigma}(K^1_{j,\sigma_1},..,K^m_{j,\sigma_m})$$

Given any $(\sigma;\sigma_1,..,\sigma_m) \in \{0,1\}^{m+1}$ such that $(\sigma;\sigma_1,..,\sigma_m) \neq (0;0,..,0)$ se can estimate

$$\| L_{j,\sigma}(K^1_{j,\sigma_1},..,K^m_{j,\sigma_m}) \| \leq \| L_{j,\sigma} \| \ \| K^1_{j,\sigma_1} \| .. \| K^m_{j,\sigma_m} \|$$

where the term in the right hand side of the inequality is majorized by

$$\| L_j - L \| \mu^m \quad \text{if} \quad \sigma = 1 \quad \text{or by} \quad \| K^r_j - K^r \| \mu^m \quad \text{if} \quad \sigma_r = 1$$

Thus, in any case we have

$$\| L_{j,\sigma}(K^1_{j,\sigma},..,K^m_{j,\sigma_m}) \| \to 0 \quad \text{if} \quad (\sigma;\sigma_1,..,\sigma_m) \neq (0;0,..,0)$$

Therefore $L_j(K^1_j,..,K^m_j) \to L(K^1,..,K^m)$ since

$$L_{j,0}(K^1_{j,0},..,K^m_{j,0}) = L(K^1,..,K^m).$$

<u>1.10. THEOREM.</u> *Let* $(f_j)_{j\in J}$ *and* $(g_j)_{j\in J}$ *be nets in* Hol(D,D_1) *and* Hol(D_1,D_2) *such that*

$$\mathcal{T}\lim_{j\in J} f_j = f \qquad \mathcal{T}\lim_{j\in J} g_j = g$$

Then we have $\mathcal{T}\lim_{j\in J} g_j f_j = gf$ *in* Hol(D,D_2).

Proof: Let us fix $a \in D$ arbitrarily. By corollary 1.7 it suffices to show that $(g_j f_j)^{(m}_a \to (gf)^{(m}_a$ for all $m \in \mathbb{N}$. Observe that from proposition 1.1 it follows that for fixed m, the mappings $x \to g^{(m}_{j,x}$ are equilipschitzian in some neighbourhood of the point $b =: f(a)$. Explicitly , we have

$$\| g_{j,x}^{(m} - g_{j,y}^{(m} \| \leq (m+1)\left(\frac{e}{\varepsilon/2}\right)^{m+1} \| x-y \|$$

if $\varepsilon =: \operatorname{dist}(b, \partial D_1)$ and $x, y \in B =: \{z \in D_1;\ \| z-b \| < \varepsilon/2\}$. Since $f_j(a) \to b$, there exists and index j_0 such that $f_j(a) \in B$ for all $j \geq j_0$. But then

$$(1.6)\qquad \| g_{j,f_j(a)}^{(m} - g_{f(a)}^{(m} \| \leq \| g_{j,f_j(a)}^{(m} - g_{j,f(a)}^{(m} \| +$$

$$+ \| g_{j,f(a)}^{(m} - g_{f(a)}^{(m} \| \leq (m+1)\left(\frac{e}{\varepsilon/2}\right)^{m+1} \| f_j(a) - f(a) \| +$$

$$+ \| g_{j,b}^{(m} - g_b^{(m} \| \to 0$$

because of theorem 1.5 (applied to g_j and the point b) and the fact that $f_j(a) \to f(a)$. Therefore, by lemma 1.9 we have

$$(g_j f_j)_a^{(k} = \sum_{m=1}^{k} \sum_{\substack{\nu_1+\dots+\nu_m=k \\ \nu_1,\dots\nu_m \geq 1}} g_{f_j(a)}^{(m} [f_a^{(\nu_1}, \dots, f_a^{(\nu_m}] \to$$

$$\to \sum_{m=1}^{k} \sum_{\substack{\nu_1+\dots+\nu_m=k \\ \nu_1,\dots\nu_m \geq 1}} g_{f(a)}^{(m} [f_a^{(\nu_1}; \dots; f_a^{(\nu_m}] = (gf)_a^{(m}.$$

#

It turns out from the proof of (1.6) that, in general, we have

<u>1.11. COROLLARY</u>. *If* $g_{j,b_j}^{(k} \to g_b^{(k}$ *and* $b_j \to b$ *then* $g_{j,b_j}^{(k} \to g_b^{(k}$. *In particular,* $g_{j,b_j}^{(k} \to g_b^{(k}$ *whenever* $b_j \to b$ *and* $T\lim_{j \in J} g_j = g$.

<u>1.12. EXERCISE</u>. (a) Prove the theorem directly by showing that given any $a \in D$, we have $\| g_j f_j - gf \|_B \to 0$ for a sufficiently small ball $B \subset\subset D$ centered at a.

(b) Let $U \subset\subset D$ and $k \in \mathbb{N}$ be given; prove that the family $f^{(k} =: x \to f_x^{(k}$, $x \in U$, $f \in \operatorname{Hol}(D, D_1)$, consists of holomorphic mappings and is uniformly bounded. What is $(f^{(k})_a^{(m}$?.

§3.- Differentiability of the composition operation.

For later purposes, it is useful to investigate also the differentiability properties of the composition operation.

1.13. DEFINITION. *Let* $I \subset \mathbb{R}$ *be an interval and suppose that for every* $t \in I$ *we are given a mapping* $f^t \in Hol(D,D_1)$. *We shall say that* $t \to f^t$ *is "T-derivable" if we have*

$$T \lim_{h \to 0} \frac{1}{h} (f^{t+h} - f^t) = A$$

for some $A \in Hol(D,E_1)$.

At this point if is a hard question whether the mapping $t \to f^t g^t$ is derivable in the T sense if so are $t \to f^t$ and $t \to g^t$. However, it is easy to prove a slightly weaker statement.

1.14. DEFINITION. *A net* $(f_j)_{j \in J}$ *in* $Hol(D,D_1)$ *is said to be "weakly T-convergent" to* f *if, for every* $x \in D$, *there exists a ball* B *centered at* x *such that* $\| f_j - f \|_B \to 0$. *We write* $T_w \lim_{j \in J} f_j = f$ *in that case.*

A map $t \to f^t$ *from* I *into* $Hol(D,D_1)$ *is said to be "weakly T-derivable" at* $t_0 \in I$ *if we have*

$$T_w \lim_{h \to 0} \frac{1}{h} (f^{t_0+h} - f^{t_0}) = A$$

for some $A \in Hol(D,E_1)$.

1.15. LEMMA. *If the maps* $I \to Hol(D,D_1)$ *and* $I \to Hol(D_1,D_2)$ *given respectively by* $t \to f^t$ *and* $t \to g^t$ *are weakly T-derivable at* t_0, *then* $t \to g^t f^t$ *is weakly T-derivable at* t_0.

Proof: We may assume $t_0 = 0$. Let us fix $x \in D$. We can choose a ball $B_1 \subset\subset D_1$ centered at $f^0 x$ such that $L =: \frac{d}{dt}\big|_0 g^t$ is bounded when restricted to B_1 and

$$\| \frac{1}{t} (g^t - g^0) - L \|_{B_1} \to 0$$

Let $B_2 \subset E_2$ be a ball centered at 0 such that $LB_1 \subset\subset B_2$. Furthermore, let $B \subset\subset D$ be a ball centered at x such that $f^0(B) \subset\subset B_1$ and

$$\| \frac{1}{t} (f^t - f^0) - K \|_B \to 0$$

where $K=:\frac{d}{dt}\big|_0 f^t$. Then, for sufficiently small values of t we have

$$f^t(B)\subset B_1 \quad \text{and} \quad \frac{1}{t}(g^t-g^0)(B_1)\subset B_2$$

Moreover

$$\frac{1}{t}(g^t f^t-g^0 f^0)=[\frac{1}{t}(g^t-g^0)]f^t+\frac{1}{t}(g^0 f^t-g^0 f^0)$$

and applying proposition 1.1 we obtain

$$\mathcal{T}\lim_{t\to 0}\frac{1}{t}(g^t-g^0)f^t\big|_B = Lf^0\big|_B$$

On the other hand, for $y\in B$ we have

$$\| g^0(f^t y)-g^0(f^0 y+tKy)\| \leqslant \| g^{0(1}\|_{B_1} \| f^t y-f^0 y-tKy\|$$

Hence

$$\|\frac{1}{t}(g^0 f-g^0 f)-\frac{1}{t}[g^0(f^0+tK')-g^0 f^0]\|_B \to 0$$

From the Taylor series of g^0 at $f^0 y$ we see that

$$\|\frac{1}{t}[g^0(f^0 y+tKy)-g^0(f^0 y)]-g^{(1}_{f^0 y}Ky\| \leqslant$$

$$\leqslant \sum_{k=2}^{\infty} t^{k-1}\| g^{(k}_{f^0 y}(Ky,\ldots,Ky)\| \leqslant$$

$$\leqslant \sum_{k=2}^{\infty} t^{k-1}\|g^{(k}\|_{B_1}\|K\|_B^k \to 0$$

Therefore

$$\frac{1}{t}(g^t f^t-g^0 f^0)\big|_B \to g^{(1}_{f^0}K\big|_B \quad \text{uniformly.}$$

#

CHAPTER 2

TOPOLOGICAL CONSEQUENCES OF THE GROUP STRUCTURE OF THE SET OF AUTOMORPHISMS

§1.- The topological group Aut D.

So far all out considerations were valid for any of the spaces $Hol(D,D_1)$. The priviledge of Aut D with respect to them is, first of all, that any mapping in it admits an inverse and that the composition operation can be iterated any number of times in it.

Henceforth we restrict our attention to Aut D. Let us consider first the continuity of the mapping $f \to f^{-1}$ with regard to $\mathcal{T}$. Since we have already established the continuity of the multiplication (i.e. the operation $(f,g) \to gf$ in Aut D, cf. Chap 1, §2), it suffices to prove that

$$(2.1) \qquad \mathcal{T}\lim_{j\in J} g_j = id_D \Rightarrow \mathcal{T}\lim_{j\in J} g_j^{-1} = id_D$$

in Aut D. Indeed, let $(f)_{j\in J}$ be a net in Aut D such that $\mathcal{T}\lim_{j\in J} f_j = f$ and write $g_j =: f_j f^{-1}$. Then we have

$$\mathcal{T}\lim_{j\in J} g_j = \mathcal{T}\lim_{j\in J} f_j f^{-1} = id_D$$

whence by assumption it follows

$$\mathcal{T}\lim_{j\in J} g_j^{-1} = \mathcal{T}\lim_{j\in J} f f_j^{-1} = id_D$$

and therefore $\mathcal{T}\lim_{j\in J} f_j^{-1} = f^{-1}$.

To see (2.1) we prove the below estimation:

2.1. LEMMA. *Let* B, B' $\subset\subset$ D *be given balls. Then there exists a constant* K *such that we have*

$$\| g^{-1}f - id_D \|_B \leq K \| f-g \|_B$$

for all f, g∈Aut D *with* $f(B)\subset B'$, $g(B)\subset B'$.

Proof: Let us fix $x\in B$ and set $a=:g(x)$. If $a, b\in B'$, then by Cauchy majorizations we have

$$\|g^{-1}f(x)-x\| = \|g^{-1}(a)-g^{-1}b\| \leqslant$$

$$\leqslant \frac{\|g^{-1}\|_D}{\operatorname{dist}(B',\partial D)}\|a-b\| \leqslant \frac{\sup\{\|y\|;\ y\in D\}}{\operatorname{dist}(B',\partial D)}\|f-g\|_B$$

#

Hence we immediately obtain the following:

2.2. THEOREM. Aut D *is a topological group with regard to the topology T.*

Proof: It suffices to show (2.1).

Let us fix any ball $B\subset\subset D$. Set $\delta=:\operatorname{dist}(B,\partial D)$ and define $B_{\delta/2}=:\{x\in D;\ \operatorname{dist}(x,B)<\delta/2\}$. Then, from $T\lim_{j\in J} g_j = id_D$ we deduce that $g_j\to id_D$ with respect to $\|\cdot\|_B$; thus, there is a $j_0\in J$ such that

$$\|g_j-id_D\|_B<\delta/2$$

for all $j\geqslant j_0$ and, in particular, $g_j(B)\subset B_{\delta/2}$. Applying lemma 2.1 to $g=:g_j$, $f=:id_D$ and $B'=:B_{\delta/2}$ we see that

$$\|g_j^{-1}-id_D\|_B\leqslant K\|g_j^{-1}-id_D\|_B$$

for all $j\geqslant j_0$. Thus $g_j^{-1}\to id_D$ with regard to $\|\cdot\|_B$ and the result follows from theorem 1.6. #

2.3. EXERCISE. Show by examples in Aut Δ that the constant K in lemma 2.1, in general, must depend on both B and B'.

2.4. REMARK. For a better understanding of the situation in lemma 2.1 it is important to note that:

Given any ball $B\subset\subset D$, there is a constant K such that we have

$$\| f-g \|_B \leqslant K \| g^{-1}f-id_D \|_B$$

for all f,g∈Aut D.

Proof: Let $B' = B_{\delta/2}$ be defined as above, where again $\delta =: \mathrm{dist}(B,\partial D)$, and consider any x∈B. Now, if the point $y =: g^{-1}f(x)$ lies in B', then by proposition 1.4 we have

$$\| f(x)-g(x) \| = \| g(y)-g(x) \| \leqslant \frac{2}{\delta} \sup\{ \| z \| \; ; \; z \in D \} \| y-x \|$$

On the other hand, if $y \notin B'$ then obviously $\| y-x \| \geqslant \delta/2$ whence

$$(2.2) \qquad \| f(x)-g(x) \| \leqslant \mathrm{diam}(D) \leqslant \frac{2}{\delta} \mathrm{diam}(D) \| x-y \|$$

Thus, by the arbitraryness of x∈B we obtain

$$\| f-g \|_B \leqslant \frac{\delta}{2} \max\{ \sup_{z \in D} \| z \| \, , \, \mathrm{diam}(D) \} \| g^{-1}f-id_D \|_B$$

#

2.5. EXERCISE. Using the fact that constant mappings have null derivative, show that we have

$$\| f-g \|_B \leqslant \frac{2}{\delta} \mathrm{diam}(D) \| g^{-1}f-id_D \|_B$$

for all f, g∈Aut D. Hint: shift D.

§2.- Cartan's uniqueness theorem.

Next we investigate the consequences of the fact that, in Aut D, the composition can be infinitely iterated. The proof of the result we shall obtain shads the first light on how the geometry of D determines the automorphisms.

2.6. THEOREM. *Let* f, g∈Aut D. *If for some* a∈D *we have* $f_a^{(0} = g_a^{(0}$ *and* $f_a^{(1} = g_a^{(1}$ then f= g.

Proof: Let us consider the map $h = g^{-1}f$. We have $h_a^{(0} = a$ and $h_a^{(1} = id$. Therefore it suffices to prove the statement of the theorem for $h =: id_D$.

Suppose $h \neq id_D$. Then there exists some $\ell \in \mathbb{N}$ such that

$$h_a^{(2} = h_a^{(3} = \ldots = h_a^{(\ell-1} = 0 \quad \text{and} \quad h_a^{(\ell} \neq 0$$

Consider the iterated maps $h^1, h^2, \ldots$ We show by induction that

$$(2.3) \qquad (h^p)_a^{(1} = \ldots = (h^p)_a^{(\ell-1} = 0 \quad \text{and} \quad (h^p)_a^{(\ell} = p\, h_a^{(\ell}$$

for all $p \in \mathbb{N}$. Obviously

$$(h^p)_a^{(0} = a \qquad \text{and} \qquad (h^p)_a^{(1} = \mathrm{id}$$

for all $p \in \mathbb{N}$, and the assertion of (2.3) for $p = 1$ is nothing but the definition of ℓ.

Assume (2.3) holds for p. By proposition 1.8 we have

$$(h^{p+1})_a^{(k} = \sum_{m=1}^{k} \sum_{\substack{\nu_1 + \ldots + \nu_m = k \\ \nu_1, \ldots, \nu_m \geq 1}} (h^p)_a^{(m}[h_a^{(\nu_1}, \ldots, h_a^{(\nu_m}]$$

for $k = 1, 2, \ldots$ Consider the case $2 \leq k \leq \ell$. Then, from

$$(h^p)_a^{(2} = \ldots = (h^p)_a^{(k} = 0 \quad \text{and} \quad h_a^{(k} = 0 \quad \text{we derive} \quad (h^{p+1})_k^{(a} = 0$$

Let us compute $(h^{p+1})_a^{(\ell}$. From $(h^p)_a^{(2} = \ldots = (h^p)_a^{(\ell-1} = 0$ we get

$$(h^{p+1})_a^{(\ell} = (h^p)_a^{(1}[h_a^{(\ell}] + (h^p)_a^{(\ell}[h_a^{(1}, \ldots, h_a^{(1}] = h_a^{(\ell} + p\, h_a^{(\ell} = (p+1) h_a^{(\ell}$$

which proves (1.3).

Hence we have $\lim_{p \to \infty} \| (h^p)_a^{(1} \| = \infty$ which contradicts the Cauchy majorizations (Chapter 1, proposition 1.1)

$$\| (h^p)_a^{(\ell} \| \leq \left(\frac{e}{\delta} \right)^{\ell} \sup\{ \|x\| \; ; \; x \in D\}$$

for all $p \in \mathbb{N}$. In fact we have proved the following:

<u>2.7. COROLLARY</u>. *Let* $h \in \mathrm{Hol}(D)$ *be given and assume that there exists a point* $a \in D$ *for which we have* $h_a^{(0} = a$, $h_a^{(1} = \mathrm{id}$. *Then* $h = \mathrm{id}_D$.

§3.- Topological version of Cartan's uniqueness theorem.

Roughly speaking, Cartan's uniqueness theorem states that,

given a point $a\in D$, the automorphisms of D depend only on their 0-th and 1-st derivatives at the point a, i.e., any $f\in \mathrm{Aut}\, D$ is uniquely determined by the pair $f_a^{(0}$, $f_a^{(1}$. Is this correspondence continuous?. The answer is affirmative. We have the following topological version of Cartan's uniqueness theorem:

<u>2.8. THEOREM</u>. *Let* $a\in D$ *be fixed and assume that* $f, f_j \in \mathrm{Aut}\, D$, $j\in J$, satisfy $f_{j,a}^{(s}\to f_a^{(s}$ *for* $s=0,1$. *Then we have* $\mathcal{T}\lim_{j\in J} f_j = f$.

Proof: Essentially, we carry out the reasonings of the proof of Cartan's theorem in a more general setting (where the role of f is now played by f_j).

Define $h_j =: f^{-1} f_j$. The relation $h_{j,a}^{(0} = f^{-1}[f_j(a)]\to a$ is clear. Observe that also

$$h_{j,a}^{(1} = (f^{-1} f_j)_a^{(1} = (f^{-1})_{f_j(a)}^{(1}\ f_{j,a}^{(1} \to (f^{-1})_{f(a)}\ f_a^{(1} = \mathrm{id}$$

in account of proposition 1.1. Next we stablish the following axuliary stament

<u>2.9. LEMMA</u>. *If* $(h_j)_{j\in J}$ *is a net in* Hol(D) *such that* $h_{j,a}^{(0}\to a$, $h_{j,a}^{(1}\to \mathrm{id}$ *and* $h_{j,a}^{(k}\to 0$ *for* $k=2,\ldots,\ell-1$ *then we have*

$$\| (h_j^p)_a^{(\ell} - p\, h_{j,a}^{(\ell} \| \to 0$$

for all $p\in \mathbb{N}$.

An inmediate consequence we obtain

<u>2.10. COROLLARY.</u> *For any net* $(h_j)_{j\in J}\subset \mathrm{Hol}(D)$ *with* $h_{j,a}^{(0}\to a$, $h_{j,a}^{(1}\to \mathrm{id}$ *we have* $\mathcal{T}\lim_{j\in J} h_j = \mathrm{id}_D$.

Indeed: If we had $h_j \not\to \mathrm{id}_D$ then by theorem 1.5 there would be an $\ell\geq 2$ such that $h_{j,a}^{(k}\to 0$ for $2\leq k\leq \ell$ and

$$\lambda =: \overline{\lim_{j\in J}}\ \| h_{j,a}^{(\ell} \| > 0$$

Then, by lemma 2.9

$$\|(h_j^p)_a^{(\ell}\| \geqslant p\|h_{j,a}^{(\ell}\| - \|(h_j^p)_a^{(\ell} - p\, h_{j,a}^{(\ell}\|$$

for all $j \in J$ and $p \in \mathbb{N}$. Therefore

$$\overline{\lim_{j \in J}}\ \|(h_j^p)_a^{(\ell}\| \geqslant p\lambda$$

for $p \in \mathbb{N}$. But this contradicts the Cauchy estimates (proposition 1.1).

$$\sup\{\|f_a^{(\ell}\|\ ;\ f \in Hol(D)\} < \infty$$

proving the corollary.

Thus, in our case $f^{-1} f_j \to id_D$ whence $\mathcal{T}\lim_{j \in J} f_j = f$.

Thus, our only remainder task is to prove the lemma. This requires a better overlook on the expansion of h^p as a direct iteration of the formula given by proposition 1.8 would lead to very involved expressions.

Instead, let us procceed as follows: Start from

$$(gf)(a+x) = g(b) + \sum_{m=1}^{\infty} \sum_{\nu_1,\ldots,\nu_m=1} g_b^{(m}\left[f_a^{(\nu_1}(x,\ldots,x);\ldots;f_a^{(\nu_m}(x,\ldots,x)\right]$$

This asserts that, for $f, g \in Hol(D)$ and for sufficiently small vectors $x \in E$, $(gf)(a+x)$ is the sum of all possible expressions

$$(2.4) \qquad g_{f(a)}^{(m}\left[f_a^{(\nu_1}(x,\ldots,x);\ldots;f_a^{(\nu_m}(x,\ldots,x)\right]$$

Let us write (2.4) in the more visualizable form

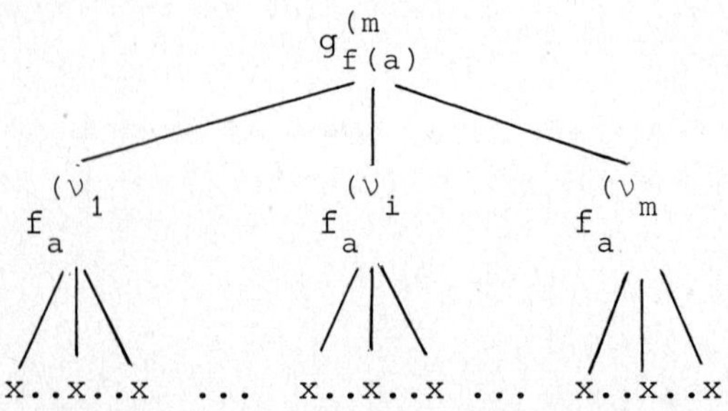

In such a way, it seems to be intuitively clear that $f^p(a+x)$ is the sum of all possible expressions corresonding to the graphs of the form

(2.5)

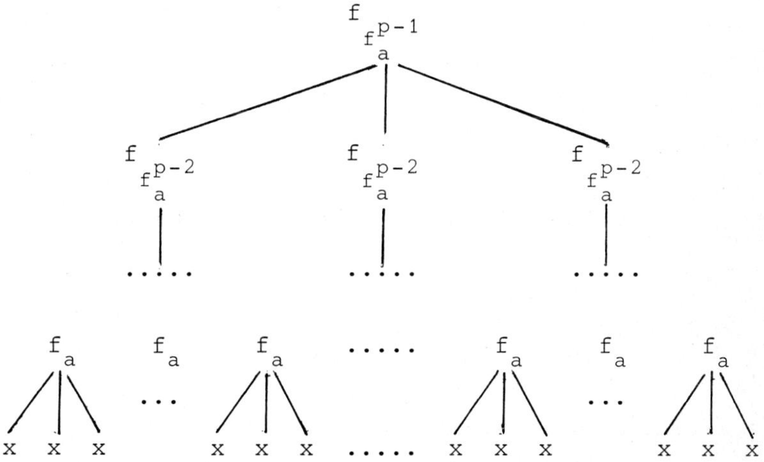

Here the symbol $\nearrow$ can be interpreted as the sign of substitution. Now we stablish the precise mathematical development of this technique. In order to be self-contained, we shall make no reference to the usual theory of tree graphs.

<u>2.11. DEFINITION</u>. *Let* $n \in \mathbb{N}$ *be arbitrarily fixed. A "tree of height n" is an* n-*uple* $A = (\alpha_0, \ldots, \alpha_{n-1})$ *of functions such that*

1) for every p, *the domain of* α_p *is a segment* dom $\alpha_p = \{1, 2, \ldots d_p(A)\}$ *of* $\mathbb{N}$.

2) range $\alpha_{n-1} = \{1\}$.

3) for $0 \leq p \leq n-1$, α_p *is a "monotone increasing" mapping "onto"* dom α_{p+1}.

4) for $p = 1, \ldots, n$ *we have* dom α_p = range α_{p-1}.

The number d (A) is called the *width of A at the height p*, and we shall write $d(A) =: d_0(A)$. We say that $d(A)$ is the *degree* of *A*.

2.12. EXAMPLE. Consider the plain graph

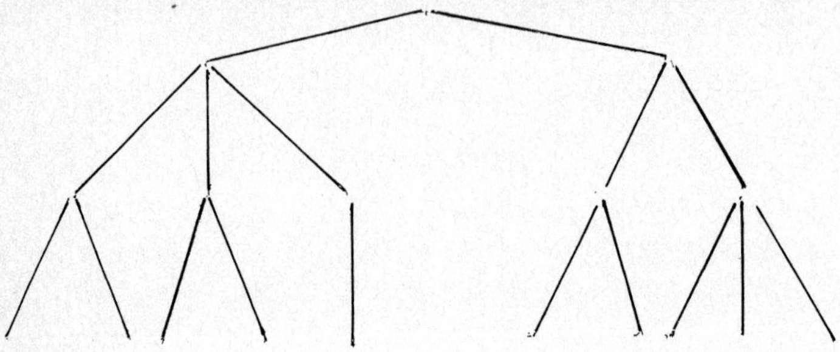

Note that the sequence of the vertices is relevant! This can be interpreted as a tree of height 3 as follows:

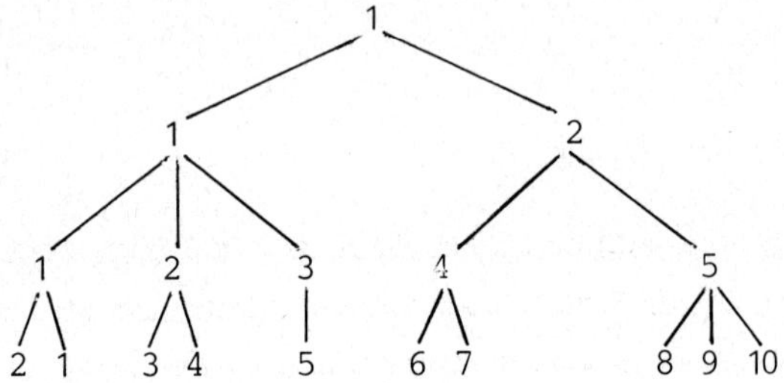

$$\alpha_1(1) = 1 \quad \alpha_1(2) = 1 \ \ldots \ \alpha_1(5) = 2$$

$$\alpha_0(1) = 1 \ldots \alpha_0(5) = 3 \ \ldots \ \alpha_0(10) = 5$$

At this point we can provide an exact interpretation of (2.5).

2.13. DEFINITION. *Let Trees* (n) *denote the set of all trees of height* n. *Given* $g \in \mathrm{Hol}(D)$ *we define the "tree-derivatives"* $g_a^{(A}$ *of* g *at* $a \in D$ *as follows:*
For $A \in$ Trees *(1) we set* $g_a^{(A} =: g_a^{(d(A)}$ *in the usual sense. If* $g_a^{(B}$ *is already defined for all* $B \in$ Trees (n-1) *and* $A = (\alpha_0, \ldots, \alpha_{n-1}) \in$ Trees (n) *we set*

$$g_a^{(A} =: g_{g(a)}^{(B}[g_a^{(\#\alpha_0^{-1}(1)}, \ldots, g_a^{(\#\alpha_0^{-1}(d(B))}]$$

where $B=(\alpha_1,..,\alpha_{n-1})\in\mathrm{Trees}(n-1)$, $\alpha_0^{-1}(k)=:\{m;\ \alpha_0(m)=k\}$ *and* $\#$ *means cardinality.*

2.14. PROPOSITION. *The series*

$$\sum_{A\in\mathrm{Trees}(n)} \| g_a^{(A}(x,..,x)\|$$

is uniformly convergente on some neighbourhood of the origin and we have

$$g^n(a+x)=g^n(a)+\sum_{A\in\mathrm{Trees}(n)} g_a^{(A}(x,..,x) \tag{2.6}$$

Proof: For n= 1 this formula is equivalent to the usual Taylor expansion of g. Remark that by the Cauchy estimates we have

$$\sum_{m=1}^{\infty} \|g_a^{(m}\|\,\rho^m \leq \sum_{m=1}^{\infty}\left(\frac{e}{\delta}\rho\right)^m$$

where $\delta=\mathrm{dist}(a,\partial D)$ and $\rho\in[0,\infty)$.

Now, assume we had proved

$$g^{n-1}(b+x)=g^{n-1}(b)+\sum_{B\in\mathrm{Trees}(n)} g_b^{(B}(x,..,x) \tag{2.7}$$

for sufficiently small vectors x and

$$\sum_{B\in\mathrm{Trees}(n-1)} \| g_b^{(B}\|\,\rho^{d(B)}\leq \omega_{g^{n-2}(a)}[\omega_{g^{n-1}(a)}(..\omega_a(\rho))] \tag{2.8}$$

for all $\rho\in[0,\infty)$, where

$$\omega_y(\rho)=\sum_{\nu=1}^{\infty}\left[\frac{e\rho}{\mathrm{dist}(y,\partial D)}\right]^{\nu}$$

We prove (2.7) and (2.8) for n. Let us begin with (2.8)

$$\sum_{A\in\mathrm{Trees}(n)} \| g_a^{(A}\|\,\rho^{d(A)}= \tag{2.9}$$

$$=\sum_{A\in\mathrm{Trees}(n)} \| g_{g(a)}^{(B}[g_a^{(\#\alpha_0^{-1}(1)},..,g_a^{(\#\alpha_0^{-1}d(B)}]\|\,\rho^{\#\alpha_0^{-1}(1)+.+\#\alpha_0^{-1}(d(B))}$$

But given any $B \in \text{Trees}(n-1)$ and $\nu_1,..,\nu_{d(B)}$ there exists a unique α_0 such that $(\alpha_0, B) \in \text{Trees}(n)$ and $\nu_k = \# \alpha_0^{-1}(k)$ for $k = 1,..,d(B)$. Thus the second member of (2.9) is

$$\sum_{B\in\text{Trees}(n-1)} \sum_{\nu_1,..,\nu_{d(B)}=1}^{\infty} \| g_{g(a)}^{(B} [g_a^{(\nu_1}, g_a^{(\nu_2} .. g_a^{(\nu_{d(B)}}] \| \rho^{\nu_1+..+\nu_{d(B)}} \leqslant$$

$$\leqslant \sum_{B\in\text{Trees}(n-1)} \sum_{\nu_1+..+\nu_{d(B)}=1}^{\infty} \| g_{g(a)}^{(B} \| \, \| g_{g(a)}^{(\nu_1} \| . \| g_{g(a)}^{(\nu_{d(B)}} \| \rho^{\nu_1+..+\nu_{d(B)}}$$

which, due to Cauchy's majorizations, is dominated by

$$\sum_{B\in\text{Trees}(n-1)} \sum_{\nu_1+..+\nu_{d(B)}}^{\infty} \| g_{g(a)}^{(B} \| \left(\frac{e\rho}{\delta} \right)^{\nu_1+..+\nu_{d(B)}} =$$

$$= \sum_{B\in\text{Tress}(n-1)} \| g_{g(a)}^{(B} \| \left[\sum_{\nu=1}^{\infty} \left(\frac{e\rho}{\delta} \right)^{\nu} \right]^{d(B)}$$

By the induction hypothesis the last sum is dominated by

$$\leqslant \omega_{g^{n-1}(a)} [\omega_{g^{n-2}(a)} (..\omega_a(\rho))]$$

Thus (2.8) is established. Now (2.8) is immediate

$$\sum_{A\in\text{Tress}(n)} g_a^{(A}(x,..x) =$$

$$= \sum_{B\in\text{Trees}(n-1)} \sum_{\nu_1,..,\nu_{d(B)}}^{\infty} g_{g(a)}^{(B} [g_a^{(\nu_1}(x,..,x); .. g_a^{(\nu_{d(B)}}(x,..,x)] =$$

$$= \sum_{B\in\text{Trees}(n-1)} g_{g(a)}^{(B} \left[\sum_{\nu_1=1}^{\infty} g_a^{(\nu_1}(x,..,x); .. \sum_{\nu_{d(B)}=1}^{\infty} g_a^{(\nu_{d(B)}}(x,..,x) \right] =$$

$$= \sum_{B\in\text{Trees}(n-1)} g_{g(a)}^{(B} [g(a+x) - g(a)] = g^{n-1}[g(a+x) - g^{n-1}(g(a))] =$$

$$= g^n(a+x) - g^n(a).$$

#

Now we can prove the lemma.

Proof of lemma 2.9: It follows from proposition 2.14 that

$$(g^p)^{(k}_a = \sum_{\substack{A\in \mathrm{Trees}(p)\\ d(A)=\ell}} g^{(A}_a \tag{2.9}$$

for any $p,k\in\mathbb{N}$, $a\in D$ and $g\in \mathrm{Hol}(D)$. Consider any $A\in\mathrm{Trees}(p)$ with $d(A)=\ell$ such that A has a vertex where the number of entering edges is different from 1 and ℓ, that is, if $A=(\alpha_0,\ldots,\alpha_{p-1})$ then there are $s,\nu\in\mathbb{N}$ with $\#\alpha_s^{-1}(\nu)\neq 1$ and $\#\alpha_s^{-1}(\nu)\neq 1$. We show that, in that case, $h^{(A}_{j,a}\to 0$. Indeed, it is easy to see that

$$\| h^{A}_{j,a}\| \leqslant \tag{2.10}$$

$$\leqslant \| h^{(\#\alpha^{-1}_{n-1}(1))}_{j,h^{n-1}_j(a)}\| \prod_{\nu_{k-2}=1}^{d(A)} \| h^{(\#\alpha^{-1}_{n-2}(\nu_{k-2}))}_{j,h^{n-2}(a)}\| \cdots \prod_{\nu_1=1}^{d(A)} \| h^{(\#\alpha^{-1}_{1}(\nu_{1}))}_{j,h(a)}\|$$

that is, $\| h^{d(A}_{j,a}\|$ is not greater than the product of the norms of all those derivatives that occur at some vertex of A if we draw $h^{(A}_{j,a}$ as

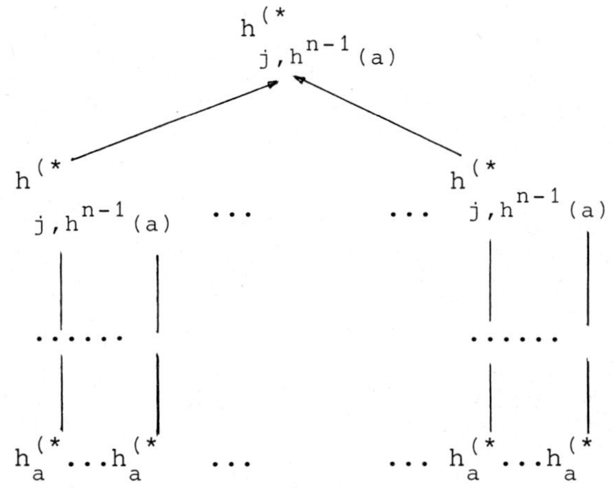

By Cauchy majorizations all the factors of (2.10) are bounded by a constant independent of j. At the same time, by corollary 1.11 we have

$$h_{j,h_j^s(a)}^{(\#\alpha_s^{-1}(\nu)} \to 0$$

since $k=\#\alpha_s^{-1}(\nu)<\ell$ and $k\neq 1$ entails $h_{j,a}^{(k}\to 0 = id_a^{(k}$ and hence $h_{j,h_j^s(a)}^{(k}\to 0$ because $h_{j(a)}^{s}\to a$. The only trees A in Trees(p) with the properties that each of their vertices admit a 1 or ℓ entering edge and $d(A)=\ell$ are

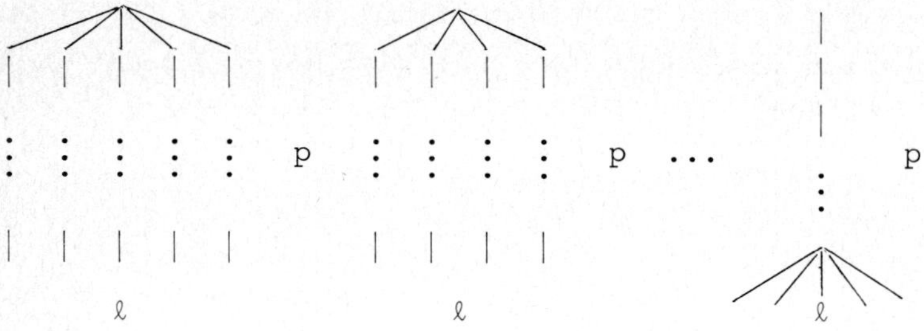

Let us call them $A_1,\ldots,A_p$. Observe that

$$h_{j,a}^{(A_\nu} = h_{j,h_j^{\ell-\nu}(a)}^{(\ell} \qquad \text{for} \quad \nu = 1,2,..p$$

Therefore $h_{j,a}^{(A_\nu}\to h_a^{(\ell}$ for $\nu = 1,..p$, whence

$$\lim_{j\in J}\left[(h_{j,a}^{p})^{(\ell}-ph_a^{(\ell}\right] = \lim_{j\in J}\left[(h_{j,a}^{p})^{(\ell}-\sum_{\nu=1}^{p} h_a^{(A_\nu}\right] =$$

$$= \lim_{j\in J} \sum_{\substack{A\in \mathrm{Trees}(p),\, d(A)=p \\ A\neq A_1,\ldots,A_p}} h_{j,a}^{(A} = 0$$

#

<u>2.15. EXERCISE.</u> (a) Write the series of $f_n f_{n-1}..f_1$ in the graph form.

(b) Prove (2.10).

CHAPTER 3

THE CARATHEODORY DISTANCE AND COMPLETENESS PROPERTIES OF THE GROUP OF AUTOMORPHISMS

§1.- The Poincaré distance.

3.1. DEFINITION. *We say that a metric* d *on a bounded domain* D *is* Aut D-*invariant if we have*

$$d(f(x),\ f(y)) = d(x,y)$$

for all $x,\ y\in D$ *and all* $f\in \mathrm{Aut}\ D$.

3.2. LEMMA. *The function*

$$d_\Delta(\zeta_1,\zeta_2) =: \left| \frac{\zeta_1-\zeta_2}{1-\bar{\zeta}_2\zeta_1} \right| , \qquad \zeta_1,\zeta_2\in\Delta$$

defines an Aut Δ-*invariant metric on* Δ.

As usually, we write $\tanh(\zeta) =: \dfrac{e^\zeta-e^{-\zeta}}{e^\zeta+e^{-\zeta}}$, $\zeta\in\mathbb{C}$, and $\tanh^{-1}$ stands for its inverse function.

Proof: Let us consider the mappings

$$U^t:\ \zeta\to \frac{\zeta+\tanh(t)}{1+\zeta\tanh(t)} , \quad \zeta\in\Delta , \quad t\in\mathbb{R}$$

It is straightforward to check that $T\lim_{j\to\infty} U^{t_j} = U^t$ for $t_j\to t$ and that $U^{t+s} = U^t\, U^s$ for all $t,\ s\in\mathbb{R}$. That is, the mapping $\mathbb{R}\to\mathrm{Aut}\ \Delta$ given by $t\to U^t$ is a continuous homomorphism. We have $d_\Delta(0,U^t(0)) = t$ for all $t\geq 0$.

We introduce the expression

$$d'_\Delta(\alpha\ \beta) =:$$

$=: \inf\{\sum_1^m t_k;\ t_k\geq 0,\ \exists\zeta_0,\ldots,\zeta_m\ ,\ \exists M_1,\ldots,M_m\in\mathrm{Aut}\ \Delta$ such that

$\zeta_0 = \alpha\ ,\ \zeta_m = \beta\ ,\ M_k(0) = \zeta_{k-1}\ ,\ M_k U^{t_k}(0) = \zeta_k$ for $k = 1..m\}$.

Trivially d'_Δ satisfies the triangle inequality and is Aut Δ-invariant.

We show that $d'_\Delta = d_\Delta$. One key observation is that

(3.1) $\quad d_\Delta(\alpha,\beta) = \inf\{t_1 \geq 0;\ \exists M \in \text{Aut}\ \Delta\ \ M(0) = \alpha,\ \ MU^{t_1}(0) = \beta\}$

because the set whose infimum is taken consists only of one element. Indeed, if $M(0) = \alpha$ and $M(\rho) = \beta$ where $\rho \in [0,1)$, then $M_{1,-\alpha}M(0) = 0$ whence, by the Schwarz lemma, there exists $k \in \partial\Delta$ such that $M_{1,-\alpha}M = M_{k,0}$. Therefore

$$k\rho = M_{k,0}(\rho) = M_{1,-\alpha}M(\rho) = M_{1,-\alpha}(\beta)$$

Thus

$$k = \frac{M_{1,-\alpha}(\beta)}{|M_{1,-\alpha}(\beta)|} \quad \text{and} \quad M = M_{1,\alpha}\, M_{k,0}$$

with this unique value of k. Then

$$\tanh(t_1) = U^{t_1}(0) = M^{-1}(\beta) = M_{\bar{k},0}\, M_{1,-\alpha}(\beta) =$$

$$= \bar{k}\, M_{1,-\alpha}(\beta) = \overline{M_{1,-\alpha}(\beta)}\ \frac{M_{1,-\alpha}(\beta)}{|M_{1,-\alpha}(\beta)|} = |M_{1,-\alpha}(\beta)| = \left|\ \frac{\beta-\alpha}{1-\bar{\alpha}\beta}\ \right|$$

so that $t_1 = d_\Delta(\alpha,\beta)$.

From (3.1) it readily follows that $d'_\Delta \leq d_\Delta$. Let us define n=:

$$=: \min\{m; \exists\alpha,\ \beta \in \Delta,\ \exists t_1,..,t_m \geq 0,\ \exists \zeta_0,..,\zeta_m,\ \exists M_1,..,M_m \in \text{Aut}\ \Delta\ ,$$

$$\text{su that}\quad \zeta_0 = \alpha,\quad \zeta_m = \beta,\ M_k(0) = \zeta_{k-1},\ M_k U^{t_k}(0) = \zeta_k \quad \text{for } k=1,..,m$$

$$\text{and}\quad t_1 + .. + t_m < d_\Delta(\alpha,\beta)\}$$

Choose α,β so that

$$\begin{array}{ll} \alpha = \zeta_0 = M_1(0) & \zeta_1 = M_1 U^{t_1}(0) \\ \zeta_1 = M_2(0) & \zeta_2 = M_2 U^{t_2}(0) \\ \ldots\ldots\ldots & \ldots\ldots\ldots\ldots \\ \zeta_{n-1} = M_n(0) & \beta = \zeta_n = M_n U^{t_n}(0). \end{array}$$

for some $M_1,..,M_n \in \mathrm{Aut}\,\Delta$, $t_1,..,t_n \geq 0$ and $\zeta_0,\zeta_1,..,\zeta_n \in \Delta$. By the definition of n we have

$$t_1+..+t_{n-1} \geq d_\Delta(\alpha,\zeta_{n-1}) = \{t\geq 0:\ M^* U^t(0) = \zeta_{n-1}\}$$

where

$$M^* =: M_{1,\alpha}\, M_{k,0} \quad \text{with } k =: \frac{M_{1,-\alpha}(\zeta_{n-1})}{|M_{1,-\alpha}(\zeta_{n-1})|}$$

Thus if $d'_\Delta \neq d_\Delta$, the only possible value of n is n=2 since there is the chain

$$\alpha = \zeta_0 = M^*(0), \quad \zeta_{n-1} = M^* U^{d_\Delta(\alpha,\zeta_{n-1})}(0) = M_n(0), \quad \beta = \zeta_n = M_n U^{t_n}(0).$$

Then we can write

$$\begin{array}{ll} M(0) = \alpha & MU^t(0) = \beta \\ M_1(0) = \alpha & M_1 U^{t_1}(0) = \zeta \\ M_2(0) = \zeta & M_2 U^{t_2}(0) = \beta \end{array}$$

By the Schwarz lemma we have $M_1 = MM_{k,0}$ for some k since $M^{-1}\, M_1(0) = M^{-1}(\alpha) = M^{-1}\, M(0) = 0$. Then, again by the Schwarz lemma, we have

$$M_2 = M_1 U^{t_1} M_{k',0}$$

$$MU^t M_{k''} = MM_{k,0} U^{t_1} M_{k',0} U^{t_2}$$

$$U^t(0) = |U^{t_1} M_{k',0} U^{t_2}(0)|$$

for suitable $k', k'' \in \partial\Delta$. However,

$$(3.2) \qquad |U^{t_1}(\lambda)| \leq U^{t_1}(|\lambda|)$$

for all $\lambda \in \Delta$ since the transformation U^{t_1} is circle-preserving and, for given $\rho > 0$, it shifts the circle $\{\lambda:\ |\lambda| = \rho\}$ in the direction of the positive real axis. Therefore

$$U^t(0) \leq U^{t_1}(|M_{k',0} U^{t_1}(0)|) = U^{t_1} U^{t_2}(0) = U^{t_1+t_2}(0)$$

whence $t \leq t_1 + t_2$.

#

3.3. EXERCISE. (a) If $M(\alpha)=N(\alpha)$, then there exists $k\in\partial\Delta$ such that $M\,M_{1,\alpha}M_{k,0}M_{1,-\alpha}=N$.

(b) Give a detailed proof for (3.2).

(c) Prove that all Aut Δ-invariant metrics are of the form

$$d(\zeta_1,\zeta_2)=f\left(\left|\frac{\zeta_1-\zeta_2}{1-\bar{\zeta}_2\zeta_1}\right|\right)$$

where $f:[0,1)\to\mathbb{R}^+$ is a function that satisfies

$$|f(\beta)-f(\alpha)|\leq f\left(\left|\frac{\beta-\alpha}{1-\bar{\alpha}\beta}\right|\right) \qquad (*)$$

for all $\alpha,\beta\in[0,1)$.

(d) Does any $f:[0,1)\to\mathbb{R}^+$ with the property (*) define an Aut Δ-invariant metric on Δ?.

3.4. REMARK. The proof of the lemma has the following heuristical background: If Δ is considered as the Poincaré modell of the Bolyai-Lobatchewsky plane (thus the straight lines are realized as circles orthogonal to $\partial\Delta$ and the group Aut Δ represents the group of even congruences of the plane), then U^t is a shift of t units in the right direction on the line $(-1,1)$.

The distance d_Δ is called the *Poincaré metric*.

§2.- The Carathéodory pseudometric.

Now we are going to construct an Aut D-invariant metric in bounded domains D of a Banach space E.

3.5. DEFINITION. *Let* D *be a bounded domain in* E. *For* $x,y\in D$ *we set*

$$(3.3)\qquad d_D(x,y)=:\sup\{d_\Delta[\psi(x),\psi(y)];\ \psi\in\mathrm{Hol}(D,\Delta)\}$$

We call d_D *the "Carathéodory pseudometric" on* D.

3.6. THEOREM. *Let* D *and* D_1 *be bounded domains in* E *and* E_1. *Then any* $f\in\mathrm{Hol}(D,D_1)$ *is a contraction with regard to the corresponding Carathéodory pseudometrics* d_D, d_{D_1}. *In particular* d_D *is an* Aut D-*invariant pseudometric.*

Proof: Given any mapping $\psi \in \mathrm{Hol}(D,\Delta)$, the expression

$$d^{\psi}(x,y) =: d_{\Delta}(\psi(x), \psi(y))$$

is trivially a pseudometric on D since d_{Δ} is a metric on Δ. Since the supremum of any family of pseudometrics is also a pseudometric, d_D is a pseudometric on D. If $f \in \mathrm{Hol}(D,D_1)$ then, for x, $y \in D$, we have

$$d_{D_1}[f(x),f(y)] = \sup\{d_{\Delta}[\psi f(x), \psi f(y)];\ \psi \in \mathrm{Hol}(D_1,\Delta)\} \leqslant$$

$$\leqslant \sup\{d_{\Delta}[\phi(x), \phi(y)];\ \phi \in \mathrm{Hol}(D,\Delta)\} = d_D(x,y)$$

If $f \in \mathrm{Aut}\, D$, then both f and f^{-1} are contractions, thus f is an isometry.

3.7. EXERCISE. Show that for $D =: \Delta$, the Carathéodory pseudometric coincides with the Poincaré metric on Δ.

§3.- The Carathéodory differential pseudometric.

Next we compare the pseudometric d_D with the natural metric induced on D by the norm of the space E. Before doing it we introduce an easily manageable relative of d_{Δ}.

3.8. DEFINITION. *For* $x \in D$ *and* $v \in E$ *we set*

$$\delta_D(x,v) =: \sup\{|\phi_x^{(1}(v)|;\ \phi \in \mathrm{Hol}(D,\Delta),\ \phi(x) = 0\} \tag{3.4}$$

We call δ_D *the "Carathéodory differential pseudometric" on* D. This terminology is explained by the following lemma:

3.9. LEMMA. (a) *The mapping* $(x,v) \to \delta_D(x,v)$, $x \in D$, $v \in E$, *is lower semicontinuous and, for any fixed* $x \in D$, $\delta_D(x,\cdot)$ *is a seminorm on* E.

(b) *If* x, $y \in D$ *and* $t \to x_t$ *is a smooth curve joining* x *with* y *so that* $x_0 = x$, $x_1 = y$, *then*

$$d_D(x,y) \leqslant \int_0^1 \delta_D(x_t, x_t')\,dt$$

Proof: The first statement follows directly from the fact that, for any fixed $\psi \in Hol(D,\Delta)$, $\delta_D^{\psi}(x,v) =: |\psi_x^{(1}(v)|$ has the mentioned properties.

Given $\phi \in Hol(D,\Delta)$, we have

$$d_\Delta[\phi(y),\phi(x)] = \int_0^1 \frac{d}{ds}\Big|_0 d_\Delta[\phi(x_{t+s}),\phi(x)]\,dt$$

which, by the triangle inequality, is dominated by

$$\int_0^1 \Big| \frac{d}{ds}\Big|_0 d_\Delta[\phi(x_{t+s}),\phi(x_t)] \Big| dt = \int_0^1 \Big| \frac{d}{ds}\Big|_0 \tanh^{-1} \frac{\phi(x_{t+s})-\phi(x_t)}{1-\phi(x_t)-\phi(x_{t+s})} \Big|$$

and, by setting $\psi_t =: M_{1,-\phi(x_t)}\phi$, the last integral can be written in the form

$$\int_0^1 |\psi'(x_t)x_t'|\,dt \leqslant \int_0^1 \delta_D(x_t,x_t')\,dt$$

since $\psi_t \in Hol(D,\Delta)$ and $\psi_t(x_t) = 0$ for all $t \in (0,1)$. #

The holomorphic mappings have the following contractive character in terms of the Carathéodory differential metric:

<u>3.10. PROPOSITION</u>. *If* $f: D \to D_1$ *is a holomorphic mapping, then we have*

$$\delta_{D_1}[f(x),\ f_x^{(1}(v)] \leqslant \delta_D(x,v)$$

for all $x \in D$ *and* $v \in E$.

Proof: We have

$$\delta_D(x,v) = \sup\{|\phi_x^{(1}(v)|;\ \phi \in Hol(D,\Delta),\ \phi(x) = 0\} \geqslant$$

$$\geqslant \sup\{|(\psi f)_x^{(1}v|;\ \psi \in Hol(D_1,\Delta),\ \psi f(x) = 0\} =$$

$$= \sup\{|\psi^1_{f(x)} f_x^{(1}v|;\ \psi \in Hol(D_1,\Delta),\ \psi f(x) = 0\} =$$

$$= \delta_{D_1}\left[f(x),\ f_x^{(1}(v)\right].$$

#

3.11. COROLLARY. *If* f: $D \to D_1$ *is biholomorphic and* $f(D) = D_1$ *then we have*

$$\delta_{D_1}\left[f(x),\ f_x^{(1}(v)\right] = \delta_D(x,v)$$

for all $x \in D$ *and* $v \in V$.

Proof:

$$\delta_D(x,v) = \delta_D\left[f^{-1}f(x),\ (f^{-1})_{f(x)}^{(1}\ f_x^{(1}\ v\right] \leq \delta_{D_1}\left[f(x),\ f_x^{(1}\ (v)\right] \leq \delta_D(x,v).$$

#

§4.- Relations between the Carathéodory pseudometric and the norm metric on D.

Now we are going to discuss the relations between the Carathéodory pseudometric and the metric induced on D by the norm of the space E.

3.12. PROPOSITION. *If* B *is the ball of radius* ρ *with center* $a \in D$, *we have*

$$\delta_B(a,v) = \frac{1}{\rho}\,\|v\| \tag{3.5}$$

and

$$d_B(a,x) = d_\Delta\left(\frac{1}{\rho}\,\|x-a\|,0\right) = \tanh^{-1}\left(\frac{1}{\rho}\,\|x-a\|\right) \tag{3.6}$$

for all $x \in B$ *and* $v \in V$.

Proof: Let $v \in E$ and $\psi \in Hol(B,\Delta)$ with $\psi(a) = 0$ be given and consider the function ϕ: $\Delta \to \Delta$ defined by $\phi(\delta) = \psi(a+\zeta v)$. We have $\phi(0) = 0$; thus, by the Schwarz lemma we have $|\phi(\zeta)| \leq |\zeta|$ for all $\zeta \in \Delta$, whence we obtain

$$\left|\psi_a^{(1}\left(\frac{\rho v}{\|v\|}\right)\right| = |\phi'(0)| \leq 1$$

Therefore $\delta_B(a,v) \leq \frac{1}{\rho}\,\|v\|$. On the other hand, given any $v \in E\setminus\{0\}$, by the Hahn-Banach theorem there exists a continuous linear functional $\psi_0 \in E^*$ such that

$$\|\psi_0\| = 1 \quad \text{and} \quad \psi_0\left(\frac{v}{\|v\|}\right) = 1$$

Then, for the function $\tilde{\psi}(x) =: \left[\frac{1}{\rho}(x-a)\right]$ we have $\tilde{\psi}(a) = 0$, $\tilde{\psi}(B) = \Delta$,

$$\tilde{\psi}_a^{(1}(v) = \lim_{s\to 0} \psi_0\left[\frac{1}{\rho s}(a+sv-a)\right] = \psi_0\left(\frac{v}{\rho}\right) = \frac{1}{\rho}\|v\|$$

This proves (3.5). To show (3.6), fix $x \in B$, set $v =: x-a$ and introduce the same functions ψ_0, $\tilde{\psi}$ as before. Since $\tilde{\psi} \in \mathrm{Hol}(B,\Delta)$, by theorem 3.6 we have

$$d_B(x,a) \leq d_\Delta|\tilde{\psi}(x), \tilde{\psi}(a)| = d_\Delta\left(\frac{1}{\rho}\|x-a\|, 0\right)$$

On the other hand, by the definition of d_B, $d_\Delta(x,a) \geq d_\Delta|\tilde{\psi}(x), \tilde{\psi}(a)|$.

3.13. THEOREM. *Let* $K \subset\subset D$ *be a convex set with* $\rho =: \mathrm{dist}(K, \partial D)$. *Then we have* $d_{D|K} \sim \mathrm{dist}\|\cdot\|_{|K}$, *or more exactly*

$$\frac{1}{\mathrm{diam}(D)}\|x-y\| \leq d_D(x,y) \leq \frac{1}{\rho}\|x-y\|$$

for all $x, y \in K$.

Proof: Given $y \in K$, let B denote the ball of center y and radius $\mathrm{diam}(D)$. Since $B \supset D$, we have $d_{B|D} \leq d_D$. But

$$d_B(x,y) = \tanh^{-1}\left[\frac{\|x-y\|}{\mathrm{diam}(D)}\right] \geq \frac{\|x-y\|}{\mathrm{diam}(D)}$$

for all $x \in B$ because the function $\tanh^{-1}$ is convex on $[0,\infty)$ and its derivative at the point 0 is equal to 1, hence $\tanh^{-1}\xi \geq \xi$ for all $\xi \geq 0$. This proves the left-hand side inequality.

To see the second inequality, fix $x, y \in K$ arbitrarily and set $x_t =: x + t(y-x)$, $B_t =:$ the ball with center x_t and radius ρ. We have $B_t \subset D$ for all $t \in [0,1]$. Therefore

$$\delta_D(x_t, v) \leq \delta_{B_t}(x_t, v) = \frac{1}{\rho}\|v\|$$

for all $v\in E$. Hence

$$d_D(x,y)\leq\int_0^1 \delta_D(x_t,x_t')\,dt\leq\int_0^1 \frac{1}{\rho}\,||x-y||\,dt=\frac{1}{\rho}\,||x-y||\,. \qquad \#$$

3.14. COROLLARY. *The topologies induced on* D *by* d_D *and the norm of* E *coincide. In particular* d_D *is not only a pseudometric but a metric.*

§5.- Completeness properties of the group Aut D.

3.15. THEOREM. *Let* $(f_j)_{j\in J}$ *be a T-Cauchy sequence in* Aut D *and assume that for some* $a\in D$ *we have* $f_j(a)\to b\in D$. *Then there exists* $f\in$ Aut D *such that* $T\lim_{j\in J} f_j = f$.

Proof: Clearly we have a unique mapping $f\in\mathrm{Hol}(D,\bar{D})$ with $T\lim_{j\in J} f_j = f$. Thus, if were $f\in$ Aut D, by theorem 2.2 we would have $T\lim_{j\in J} f_j^{-1} = f^{-1}$.

We show that

(3.7) $\qquad (f_j^{-1})_{j\in J}$ is a T-Cauchy sequence.

Let us write $b_j =: f_j(a)$ and choose any balls $B\subset\subset D$ and $B'\subset\subset D$ centered respectively at a and b such that $f(B)\subset B'$. From theorem 3.13 it follows that

$$d_D[f_j(B),\, f(B)] =: \sup_{x\in B} d_D[f_j(x), f(x)]\to 0$$

Hence we can fix $\delta>0$ so that the set

$$C =: \{x\in D;\ d_D(x,a)<\delta\}$$

is contained in B. Let us introduce also

$$C' =: \{y\in D;\ d_D(y,b)<\tfrac{1}{2}\delta\}$$

We may assume without loss of generality that $d_D(b_j,b)<\frac{1}{2}\delta$ for all $j\in J$. Then, from the triangle inequality applied to d_D and theorem 3.6 we obtain

(3.8) $\qquad f_j(C) = \{y\in D;\ d_D(y,b_n)<\delta\}\supset C'$

for all $j \in J$. Let us consider any $y \in C'$. By theorem 3.6 we have

$$d_D[f_j^{-1}(y), f_k^{-1}(y)] = d_D[y, f_j f_k^{-1}(y)] =$$

$$= d_D[f_k f_k^{-1}(y), f_j f_k^{-1}(y)] \leq d_D[f_k(C), f_j(C)]$$

because $f_j^{-1}(y) \in C$. Therefore

$$\lim_{j,k} d_D[f_j^{-1}(C'), f_k^{-1}(C')] = 0$$

But then, by theorem 3.13 we also have

$$\lim_{j,k} \| f_j^{-1} - f_k^{-1} \|_{C'} = 0$$

Since by corollary 3.14, C' contains some open ball, from theorem 1.6 we obtain that

$$\lim_{j,k} \| f_j^{-1} - f_k^{-1} \|_{B_1} = 0$$

for any ball $B_1 \subset\subset D$. Thus (3.7) is established and it follows the exitence of a unique $g \in Hol(D, \bar{D})$ such that $\mathcal{T}\lim_{j \in J} f_j^{-1} = g$.
The theorem is now a consequence of (3.7) and (3.8). Indeed, (3.8) implies $f_j^{-1}(C') \subset C$ for all $j \in J$ so that $g(C') \subset \bar{C} \subset\subset D$.
Therefore the mapping fg is defined on C' and we have $fg = \lim_{j \in J} f_j f_j^{-1} = id$ on C'. To complete the proof we must show that $g(D) \subset D$ so that fg is defined on D and $fg = id_D$.
Let us fix any $x \in D$. We have already seen that, by setting $g_j' =: f_j^{-1}|_{C'}$ and $f_j' =: f_j|_C$, the relation

$$\mathcal{T}\lim_{j,k \in J} f_j' g_k' = f \circ g|_{C'}$$

holds. Hence it follows $(f_j f_k^{-1})_b^{(s} \to (id_D)_b^{(s}$ for $s = 0,1$. Theorem 2.7 implies $\mathcal{T}\lim_{j,k} f_j f_k^{-1} = id_D$. In particular, $f_j f_k^{-1}(x) \to x$ so that, by theorem 3.6

$$d_D(f_j f_k x, x) \to d_D(x,x) = 0$$

Thus given any ball $B \subset\subset D$ centered at x, there is $j_0 \in J$ such that $(f_{j_0} f_k^{-1} x)_{k \geq j_0}$ is a d_D-Cauchy net contained in B. By theorem 3.13, this means that $(f_{j_0} f_k^{-1}x)_{k\geq j_0}$ is also a Cauchy net with regard to the norm, and hence it admits a limit $y\in\bar{B}$. Therefore we have

$$g(x)= \lim_{k\in J} f_k^{-1}x= f_{j_0}^{-1}\lim_{k\in J} f_{j_0} f_k^{-1}x= f_{j_0}^{-1}y\in D.$$

Since (3.7) is true and $f_j^{-1}(b)\to a$, we may apply all our results to $(f_j^{-1})_{j\in J}$ instead of $(f_j)_{j\in J}$. Hence $gf= id_D$, that is, $f\in Aut\ D$.

#

3.16. REMARK. In general, Aut D is not $\mathcal{T}$-complete as the following example shows:

Let us put D=: Δ and define f_n=: U^n by means of

$$U^n:\ \zeta\to \frac{\zeta+\tanh(n)}{1+\zeta\tanh(n)},\qquad n\in\mathbb{N}$$

Then U^n is $\mathcal{T}$-convergent to the constant mapping U: $\zeta\to 1$; however $U\notin Aut\ \Delta$.

This phenomenon justifies the assumption of the existence of a point $a\in D$ such that $\lim_{j\in J} f_j(a)= b\in D$ in the theorem.

Let us note, furthermore, that the boundedness of D is also of crucial importance as shown by this example:

Let us take D=: E and define $f_n\in Aut\ D$ by means of

$$f_n:\ x\to \frac{1}{n}x \qquad x\in E,\quad n\in\mathbb{N}.$$

Then f_n is $\mathcal{T}$-convergent to the constant mapping f: $x\to 0$ which clearly is not in Aut E.

This second example shads some light on the use of the Carathéodory metric in the proof. Roughly speaking it serves to establish that the Fréchet derivatives of the f_j tend to operators which are bounded from below.

3.17. EXERCISES. (a) Prove that if $D_1\supset D$, then $d_{D_1}|_D\geq d_D$.

(b) Show that given $f\in Aut\ D$ and $x\in D$, we have $\delta_D(x,v)= \delta_D[f(x),\ f_x^{(1}(v)]$ for all $v\in E$.

(c) Prove that the left uniform structure of T on Aut D is complete.

(d) On Aut D, the topologies of pointwise convergence and uniform convergence on compact subsets of D coincide.

(e) Show that, for any Banach space E, the open unit ball B(E) of E is always $d_{B(E)}$-complete.

(f) Prove that the domain of $\mathbb{C}$ defined by $D=: \Delta\setminus\{0\}$ is not d_D-complete.

3.18. THEOREM. *Let* D *be a bounded homogeneous domain. Then* D *is a complete metric space with regard to* d_D.

Proof: Let $(x_j)_{j\in J}\subset D$ be a Cauchy net with regard to the metric d_D. Fix any point $a\in D$ and set $\delta=: \text{dist}(a,\partial D)$. Then, if B denotes the ball (for the metric of the norm on D) centered at a with radius $\rho<\delta$, we have $B\subset D$. By corollary 3.14 we can find a ball $\mathcal{B}$ (for the metric d_D) centered at a with some radius r such that we have

$$\mathcal{B}\subset B\subset\subset D.$$

Since $(x_j)_{j\in J}$ is a d_D-Cauchy net, we can fix an index $j_0\in J$ so that

$$d_D(x_k,x_{j_0})<r$$

for all $k\in J$, $k\geqslant j_0$. Now, as D is homogeneous, there exists some $g\in\text{Aut } D$ such that $g(x_{j_0})= a$. Thus, as d_D is Aut D-invariant, we get

$$d_D[g(x_k),a]= d_D[g(x_k),g(x_{j_0})]= d_D(x_k,x_{j_0})<r$$

so that

$$g(x_k)\in\mathcal{B}\subset B\subset\subset D$$

for all $k\in J$, $k\geqslant j_0$. By theorem 3.13, d_D and the norm metric are equivalent on B; therefore, $(g(x_k))_{k\geqslant j_0}$ is a Cauchy net for

$\|\cdot\|$ and there exists some $\zeta\in\bar{B}\subset D$ such that

$$\lim_{k\to\infty}\| g(x_k)-\zeta\| = 0 ,$$

from which we derive

$$\lim_{k\to\infty} d_D[g(x_k),\zeta] = 0$$

again by theorem 3.13. As $\zeta = g(\eta)$ for some $\eta\in D$ and d_D is Aut D-invariant, we have

$$\lim_{k\to\infty} d_D(x_k,\eta) = \lim_{k\to\infty} d_D[g(x_k),g(\eta)] = \lim_{k\to\infty} d_D[g(x_k),\zeta] = 0$$

so that D is d_D-complete. #

CHAPTER 4

THE LIE ALGEBRA OF COMPLETE VECTOR FIELDS

Perhaps the most interesting question concerning Aut D is how the automorphisms can be determined from the shape of D. To tackle this problem directly seems to be hopeless in general. It is much more fruitful, as it was first observed by E.Cartan, to begin the considerations with the description of those automorphisms that, in some sense, lie very near to the identity. The background of this idea is the expectation that, if the boundary of D is a smooth hypersurface and $f\in$Aut D is close to id_D, then, if the vector field $v=: x\to x-f(x)$ can be continuously extended to ∂D, v must be almost tangent to ∂D. Thus, intuitively, if the difference between f and id_D is infinitesimal, then $f-id_D$ is a holomorphic vector field of infinitely small vectors that are tangent to ∂D. It can be hoped that the calculation of such vector fields is easier than the solution of the original problem. However, we also have to consider the case of a non smooth boundary and the case in which the elements of Aut D cannot be continuously extended to ∂D.

§1.- One-parameter subgroups.

In what follows we shall denote by $[t]$ the entire part of the real number $t\in\mathbb{R}$.

<u>4.1. PROPOSITION</u>. *Let* $(f_j)_{j\in\mathbb{N}}\subset$Aut D, $(t_j)_{j\in\mathbb{N}}\subset\mathbb{R}_+^*$ *with* $t_j\to 0$ *and* $A\in$Hol(D,E) *be given, and assume that the sequence*

$$A_j =: \frac{1}{t_j}(f_j - id_D)$$

satisfies $T\lim_{j\to\infty} A_j = A$. *Then, for every* $t\in\mathbb{R}$, *there exists a unique* $f^t\in$Aut D *such that*

$$\mathcal{T}\lim_{j\to\infty} f_j^{[t\, t_j^{-1}]} = f^t$$

Moreover, the mapping $\mathbb{R}\to\mathrm{Aut}\ D$ *given by* $t\to f^t$ *is a T-continuous homomorphism.*

Proof: The proof presented here is certainly not the shortest, but it provides a deep insight into the background of the theory.

We define linear operators $\hat{F}_j$ and $\hat{A}_j$ on Hol(D,E) as follows:

$$(4.1) \qquad \hat{F}_j(f) =: f\circ f_j \qquad \hat{A}_j =: \frac{1}{t_j}(\hat{F}_j - \hat{I})$$

Observe that, if we write $\hat{A}f$ for the function $x\to f_x^{(1}A(x)$, $x\in D$, we have

$$(4.2) \qquad f_j = \hat{F}_j(id_D) = (\hat{I}+t_j\hat{A}_j)\, id_D$$

$$(4.3) \qquad \mathcal{T}\lim_{j\to\infty} \hat{A}_j(f) = \hat{A}(f) \quad \text{in } U$$

whenever $U\subset D$ is such that $\|A\|_U<\infty$, $\|f\|_U<\infty$.

Proof of (4.3): Let $B\subset\subset D$ be a ball and fix $\delta>0$ so that $B_\delta =: B+\delta B(E)\subset U$. Then, by the Taylor expansion of f, we have

$$\hat{A}_j f(x) = \frac{1}{t_j}\{f[f_j(x)]-f(x)\} = \frac{1}{t_j}\{f[x+t_jA_j(x)]-f(x)\} =$$

$$= f_x^{(1}[A_j(x)] + \sum_{n=2}^{\infty} t_j^{n-1} f_x^{(n}[A_j(x),\dots,A_j(x)]$$

for all $x\in B$. Using the Cauchy estimates we obtain

$$\Big\| \sum_{n=2}^{\infty} t_j^{n-1} f_x^{(n}[A_j(x),\dots,A_j(x)] \Big\|_B \leqslant t_j \sum_{n=2}^{\infty} \frac{1}{\delta^n}\|A_j\|_{B_\delta}^n \|f\|_{B_\delta} \to 0$$

Moreover, from the Cauchy estimates we derive

$$\|f^{(1}A_j - f^{(1}A\|_B = \|f^{(1}(A_j - A)\|_B \leqslant$$

$$\leq \| f^{(1} \|_B \, \| A_j - A \|_B \leq \frac{1}{\delta} \, \| f \|_{B_\delta} \, \| A_j - A \|_B \to 0$$

so that $\mathcal{T}\lim_{j\to\infty} f^{(1}A_j = f^{(1}A$ in U. The relation (4.2) is obvious. Henceforth, let us fix a ball $B \subset D$ and a number $\delta > 0$ with the property $\| A \|_{B_\delta} < \infty$. Without loss of generality we may assume that

$$M =: \sup\{ \| A_j \|_{B_\delta} \; ; \; j \in \mathbb{N} \} < \infty$$

Observe that, as any holomorphic mapping is locally bounded, for every $x \in D$ there is a ball U centered at x such that $\| A \|_U < \infty$. Let us define

$$\exp(t\hat{A})f =: \sum_{k=0}^{\infty} \frac{t^k}{k!} \hat{A}^k f$$

We shall show that there exists a number $\tau > 0$ such that, for every $t \in [0,\tau]$, we have

$$\text{(4.4)} \qquad \hat{F} f = (\hat{I} + t_j \hat{A}_j)^{[tt_j^{-1}]} f \to \exp(t\hat{A})f$$

with respect to the norm $\| \cdot \|_B$, and

$$\text{(4.5)} \qquad \sum_{k=0}^{\infty} \frac{t^k}{k!} \| \hat{A}_k id_D \|_B < \delta$$

Then, we can easily deduce the proposition from (4.4) and (4.5). Indeed, if $t \in [0,\tau]$, we have

$$f_j^{[tt_j^{-1}]}(x) = (\hat{I} + t_j \hat{A}_j)^{[tt_j^{-1}]} id_D(x) \to \sum_{k=0}^{\infty} \frac{t^k}{k!} (\hat{A}^k id_D) x$$

uniformly for $x \in B$. By (4.5)

$$\| \sum_{0}^{\infty} \frac{t^k}{k!} (\hat{A}^k id_D) x - x \| < \delta$$

so that we have

$$\sum_{k=0}^{\infty} \frac{t^k}{k!} (\hat{A}^k id_D) x \in B_\delta \subset\subset D$$

for all $x \in B$. Then theorems 3.15 and 1.6 ensure that (this is the main point of the proof!):

$$\mathcal{T}\lim_{j\to\infty} f_j^{[tt_j^{-1}]} = f^t \in \mathrm{Aut}\, D$$

for all $t\in[0,\tau]$. Now, the mapping $[0,\tau]\to\mathrm{Hol}(D,E)$ given by $t\to f^t$ is analytic in the real sense if we endow Hol(D,E) with the norm $\|\cdot\|_B$; hence, by theorem 1.6 it is also $\mathcal{T}$-continuous. In the same way as in elementary analysis we can establish the relation

$$f^{s+t} = f^s f^t \tag{4.6}$$

for all $s,t\in[0,\tau]$ with $s+t\in[0,\tau]$. Finally, for $t\in\mathbb{R}$, we extend the definition of f^t by setting

$$f^t =: (f^{\frac{1}{2}\tau})^{[2t\tau^{-1}]} f^{t-\frac{1}{2}\tau[2t\tau^{-1}]}$$

One checks in a straightforward manner that (4.6) holds for any $s,t\in\mathbb{R}$. Then, by theorem 2.2, we have

$$\mathcal{T}\lim_{j\to\infty} f^{[tt_j^{-1}]} = f^t \in \mathrm{Aut}\, D$$

for all $t\in\mathbb{R}$, and the mapping $\mathbb{R}\to\mathrm{Aut}D$ given by $t\to f^t$ is $\mathcal{T}$-continuous since it is so in $[0,\tau]$.

Proof of (4.4): Let us consider any $g\in\mathrm{Hol}(D,E)$ with $\|g\|_K<\infty$ where K is ball concentric with B and $K_{2\varepsilon} =: K+2\varepsilon B(E)\subset B_\delta$. Assume that the index j is large enough to have

$$\|f_j - \mathrm{id}_D\|_{B_\delta} < \varepsilon$$

Then, if we write $v =: f_j(x)-x$ for any $x\in K$, we have

$$\|(\hat{A}_j g)x\| = \frac{1}{t_j}\|g[f_j(x)]-g(x)\| = \frac{1}{t_j}\left\|\int_0^1 g^{(1}_{x+sv}(v)\,ds\right\| \leqslant$$

$$\leqslant \frac{1}{t_j}\|g^{(1}\|_{K_\varepsilon}\|f_j - \mathrm{id}_D\|_K$$

which, by the Cauchy estimates, is dominated by

$$\frac{1}{\varepsilon}\|g\|_{K_{2\varepsilon}}\|A_j\|_K$$

Using this observation and taking $\varepsilon =: \frac{1}{2n}\delta$ we obtain

$$\|\hat{A}_j^n f\|_B = \|\hat{A}_j(\hat{A}_j^{n-1}f)\|_B \leqslant \frac{2n}{\delta}\|A_j\|_B \|\hat{A}_j^{n-1}f\|_{B_{\delta/n}} \leqslant$$

$$\leqslant \frac{2n}{\delta}\|A_j\|_B \frac{2n}{\delta}\|A_j\|_{B_{\delta/n}} \|\hat{A}^{n-2}f\|_{B_{2\delta/n}} \leqslant \ldots \leqslant$$

$$\leqslant \frac{2n}{\delta}\|A_j\|_B \frac{2n}{\delta}\|A_j\|_{B_{\delta/n}} \cdots \frac{2n}{\delta}\|A_j\|_{B_{(n-1)\delta/n}} \|f\|_B$$

whenever $\|f_j - id_D\|_{B_\delta} < \delta/n$, i.e, if $t_j\|A_j\|_{B_\delta} < \delta/2n$. Thus

$$\|\hat{A}_j^n f\|_B \leqslant \left(\frac{2nM}{\delta}\right)^n \|f\|_{B_\delta} \tag{4.7}$$

whenever $n < \frac{\delta}{2Mt_j}$. Therefore we may apply (4.6) for estimating the binomial expansion of $(\hat{I}+t_j\hat{A}_j)^{[tt_j^{-1}]}$ for any $t \in [0, \frac{\delta}{2M})$. This is perhaps the crucial point of the whole section. Let us fix $t \in [0, \frac{\delta}{2M})$ and write $n_j =: [tt_j^{-1}]$. We have $n_j < \frac{\delta}{2Mt_j}$. We know that

$$(\hat{I}+t_j\hat{A}_j)^{n_j} f = \sum_{k=0}^{n_j} \binom{n_j}{k} t_j^k \hat{A}_j^k(f) \tag{4.8}$$

For fixed k we have

$$\binom{n_j}{k} t_j^k = \frac{1}{k!}(n_j t_j)(n_j-1)t_j \ldots (n_j-k+1)t_j \to \frac{t^k}{k!}$$

and, by (4.3),

$$\hat{A}_j^k(f) \to \hat{A}\ (f) \qquad \text{in } \|\cdot\|_B \tag{4.9}$$

Thus the proposition follows from the below elementary lemma:

<u>4.2. LEMMA</u>. *If in a Banach space* E_1, *we have* $a_j^s \to a^s$ $(s = 1,2,\ldots)$ *where* $\sum_{s=1}^{\infty}\|a^s\| < \infty$, *and there exist* $\omega_1, \omega_2, \ldots$ *such that* $\|a_j^s\| \leqslant \omega_s$ *for all* $j,s = 1,2\ldots$ *with* $\sum_{s=1}^{\infty}\omega_s < \infty$, *then* $\sum_{s=1}^{\infty} a_j^s \to \sum_{s=1}^{\infty} a^s$.

Indeed, it suffices to apply the lemma to the space $E_1 =: \{h \in Hol(B,E);\ h \text{ is bounded on } B\}$ with the norm $\|\cdot\|_B$ and to the vectors $a_j^s =: \binom{n_j}{s} t_j^s \hat{A}_j^s(f)|_B$, $a^s =: \frac{t^s}{s!}\hat{A}^s(f)|_B$ with the constants $\omega^s =: (\frac{2tM}{\delta})^s \| f \|_{B_\delta}$.

Proof of the lemma: We have $\| a^s \| \leqslant \omega^s$ for all $s \in \mathbb{N}$; thus the series $\sum_{s=1}^{\infty} a^s$ is absolutely convergent. Then

$$\| \sum_s a_j^s - \sum_s a^s \| \leqslant \sum_s \| a_j^s - a^s \| \leqslant \sum_{s=1}^{N} \| a_j^s - a^s \| + \sum_{s=N+1}^{\infty} \omega_s$$

for all $N \in \mathbb{N}$: Given $\varepsilon > 0$, there exists N such that $\sum_{s=N+1}^{\infty} \omega_s < \varepsilon/2$. Once such an N has been fixed, there exists $j_0 \in \mathbb{N}$ such that $\sum_{s=1}^{N} \| a_j^s - a^s \| < \varepsilon/2$ for all $j \geqslant j_0$. This completes the proof.

4.3. COROLLARY. *For any* $x \in D$, *the mapping* $\mathbb{R} \to D$ *given by* $t \to f^t(x)$ *satisfies*

$$\frac{d}{dt} f^t(x) = A[f^t(x)]$$

on the whole real line.

4.4. DEFINITION. *A vector field* $x \to A(x)$, $x \in D$, *is said to be "complete in D" if the maximal solution of the initial value problem*

$$\frac{d}{dt} x^t = A(x^t) \ , \quad x^t \in D \ , \quad x^0 = x \tag{4.10}$$

is defined on the whole $\mathbb{R}$.

If $A \in Hol(D,E)$ is a vector field which is complete in D, then we define

$$\exp A =: x \to x^1, \qquad x \in D$$

where x^1 is implicitely given by (4.10).

Thus, what we have proved can be reformulated as

4.5. THEOREM. *Let* $(f_j)_{j \in \mathbb{N}}$ *be a sequence in* Aut D *and assume that, for some* $A \in Hol(D,E)$ *and some* $(t_j)_{j \in \mathbb{N}} \subset \mathbb{R}_+^*$ *with* $t_j \to 0$, *we have*

$$\mathcal{T}\lim_{j \to \infty} \frac{1}{t_j} (f_j - id_D) = A$$

Then the vector field A *is complete in* D *and we have*

$$\mathcal{T}\lim_{j\to\infty} f_j^{[tt_j^{-1}]} = \exp(tA)$$

for all $t\in\mathbb{R}$. *Moreover, the mapping* $\mathbb{R}\to\mathrm{Aut}D$ *given by* $t\to\exp(tA)$ *is a* $\mathcal{T}$*-continuous homomorphism.*

4.6. EXERCISES. (a) Give a detailed proof for (4.9)

(b) Given any $A\in\mathrm{Hol}(D,E)$ and $t_0\in\mathbb{R}$, let us define $\exp(t_0A)$ as follows:

dom $\exp(t_0A) = \{x\in D$; there is an interval $J\subset\mathbb{R}$, $[0,t_0]\subset J$, for which the initial value problem (4.10) has a solution on $J\}$.

$\exp(t_0A)x =: x^{t_0}$ where x^{t_0} is implicitely defined by (4.10).

1) Check that this is an extension of the definition of exp A as previosuly stated for complete vector fields and $t_0 = 1$.

2) Show, as in the proof of (4.4), that we have $\mathcal{T}\lim_{n\to\infty}(I+\frac{t}{n}A)^n = \exp(tA)$ on each open U subset $U\subset\subset$ dom $\exp(tA)$.

3) Show that, for each open subset U with $U\subset\subset$ dom $\exp(tA)$, we have $\mathcal{T}\lim_{n\to\infty}\sum_{k=0}^{n}\frac{1}{k!}\hat{A}^k(\mathrm{id}_D) = \exp(tA)$.

4) Use lemma 2.4 to show theorem 1.6.

§2.- Complete holomorphic vector fields.

The following question raises naturally from the previous theorem: Given a $\mathcal{T}$-continuous one-parameter group $t\to f^t$ in Aut D, is it always possible to find $A\in\mathrm{Hol}(D,E)$ such that $f^t = \exp(tA)$ for all $t\in\mathbb{R}$? It is the same as asking whether the $\mathcal{T}\lim_{n\to\infty}\frac{1}{n}(f^{1/n}-\mathrm{id}_D)$ necessarily exists, or whether the $\mathcal{T}$-continuity of $t\to f^t$ always entails its $\mathcal{T}$-differentiability.

Our next task will be to prove that the answer is affirmative. This requires a careful study of the position of the points $f^{k/n}(x)$, $k = 1,2,\ldots$ in comparison with $x+x(f^{1/n}(x)-x)$ for large n. In this section, we prove a lemma due to H. Cartan which, in many aspects, can be considered as a discrete version of the Bellman lemma in the theory of ordinary differential equations.

4.7. LEMMA. *Let* $f\in \mathrm{Aut}\, D$ *be given and assume that* $B\subset\subset D$ *is a ball with* $0<d<\mathrm{dist}(B,\partial D)$. *Suppose that, for* $x\in B$, *we have* $x_0,x_1,\ldots,x_{n-1}\in B_\delta$ *where* $x_p =: f^p(x)$ *and* $\delta<d$. *Then, by setting* $y_p =: x_0+p[f(x_0)-x]$, *we have*

$$\| x_p-y_p \| \leqslant \frac{1}{d-\delta} \| x_1-x_0 \| \sum_{k=1}^{p-1} \| f^k-id_D \|_{B_d}, \quad 1\leqslant p\leqslant n$$

Proof: If $p\leqslant n-1$, by the Cauchy estimates, we have

$$\| x_{p+1}-y_{p+1} \| \leqslant \| x_p-y_p \| + \| (x_{p+1}-x_p)-(y_{p+1}-y_p) \| =$$

$$= \| x_p-y_p \| + \| [f^p(x_1)-f^p(x_0)]-(x_1-x_0) \|$$

$$= \| x_p-y_p \| + \| [f^p-id_D](x_1)-[f^p-id_D](x_0) \| \leqslant$$

$$\leqslant \| x_p-y_p \| + \frac{1}{d-\delta} \| f^p-id_D \|_{B_d} \| x_1-x_0 \|$$

Since $x_0= y_0$ and $x_1= y_1$, the statement of the lemma follows by induction

4.8. COROLLARY. *In the above conditions, we have*

$$\| (f^p-id_D)-p(f-id_D) \|_B \leqslant \frac{1}{d-\delta} \| f-id_D \|_B \sum_{k=1}^{p-1} \| f^k-id_D \|_{B_d}$$

whenever $\| f^k-id_D \|_B <\delta$ *for* $k= 1,2,..,p-1$. *In particular we have*

$$(4.11) \qquad \| (f^p-id_D)-p(f-id_D) \|_B \leqslant \frac{\delta}{d-\delta}\, p \| f-id_D \|_B$$

whenever $p\leqslant p(\delta)$, *where*

$$p(\delta) =: \max\{p: \| f^k-id_D \|_{B_d} <\delta \quad \text{for all} \quad k\leqslant p\}.$$

Proof: We have

$$x_p-y_p = [f^p(x)-x]-p[f(x)-x]$$

whence the assertions are immediately obtained by the arbitrariness of $x \in B$.

<u>4.9. REMARK.</u> Roughly speaking, the lemma and its corollary assert that the triangle x_0, x_p, y_p is almost symmetric and its angle at x_0 is infinitesimal in comparison with $\max\{ \frac{1}{d} \| f^k - id_D \|_B , k = 1,\dots,p\}$. For $v_1, v_2 \in E$, we can define

$$\text{angle}(v_1, v_2) =: 2 \sin^{-1} \left\| \frac{v_1}{2\|v_1\|} - \frac{v_2}{2\|v_2\|} \right\|$$

However, in general, this concept is not so useful as it is in the case of Hilbert spaces.

It is important to know the limits within which the inequality (4.11) can be applied. Thus, we give upper and lower bounds for $p(\delta)$.

<u>4.10. LEMMA.</u> *Let* $f \in \text{Aut } D$ *be given and assume that* $B \subset\subset D$ *is a ball with* $d < \text{dist}(B, \partial D)$. *Then the number* $p(\delta)$ *satisfies*

$$(4.12) \qquad \frac{\delta}{\| f - id_D \|_{B_{d+\delta}}} - 1 \leq p(\delta) \leq \frac{2\delta}{\| f - id_D \|_D}$$

for all δ, $0 < \delta < \frac{1}{4}d$.

Proof: Fix an arbitrary $z_0 \in B_d$ and set $z_p =: f^p(z_0)$ for $p = 0, 1, \dots$ By the definition of $p(\delta)$ we have $z_p \in B_{d+\delta}$ if $k \leq p(\delta)$. Thus, for $p \leq p(\delta) + 1$ we obtain

$$\| z_p - z_0 \| \leq \| z_0 - z_1 \| + \dots + \| z_{p-1} - z_p \| = \sum_{k=0}^{p-1} \| (f - id_D) z_k \| \leq p \| f - id_D \|_{B_{d+\delta}}$$

whence, by the arbitrariness of z_0 in B_d, we get

$$(4.13) \qquad \| f^p - id_D \|_{B_d} \leq p \| f - id_D \|_{B_{d+\delta}}$$

for $p < 1 + p(\delta)$. But in the case $p = 1 + p(\delta)$ we have

$$\delta \| f^{1+p(\delta)} - id_D \|_{B_d} \leq [1 + p(\delta)] \; \| f - id_D \|_{B_{d+\delta}}$$

To prove the second inequality we make use of (4.11). We have

$$\delta > \| f^{p(\delta)} - id_D \|_B \geq p(\delta) \| f - id_D \|_B - \frac{1}{d-\delta} \| f - id_D \|_B \sum_{k=1}^{p(\delta)-1} \| f - id_D \|_{B_d}$$

$$> \| f - id_D \|_B \, p(\delta)(1 - \frac{\delta}{d-\delta})$$

Thus, for $\delta < \frac{d}{4}$ we have

$$\delta > \| f - id_D \|_B \, p(\delta)(1 - \frac{d/4}{3d/4})$$

#

Finally, from the first inequality or corollary 4.6 and (4.13) we derive the following estimate in terms of $f-id_D$ only:

$$(4.14) \quad \| (f^p - id_D) - p(f - id_D) \|_B \leq \frac{p(p-1)}{2(d-\delta)} \| f - id_D \|_B \| f - id_D \|_{B_{d+\delta}}$$

for all $p \leq 1 + p(\delta)$.

4.11. THEOREM. *Let* $t \to f^t$ *be a* T*-continuous one-parameter group in* Aut D. *Then there exists a unique holomorphic vector field* A *in* D *such that* A *is complete in* D *and we have* $f^t = \exp(tA)$ *for all* $t \in \mathbb{R}$,

Proof: By theorem 4.5 it suffices to show that the sequence $A_n =: n(f^{1/n} - id_D)$, $n \in \mathbb{N}$, is T-convergent in Hol(D,E). Thus it suffices to prove that $(A_n)_{n \in \mathbb{N}}$ is a T-Cauchy sequence.

Let $B \subset\subset D$ be a ball and assume that $\varepsilon > 0$ has been given. Choose $\delta > 0$ small enough to have $\frac{\delta}{d-\delta} < \varepsilon$. By assumption we can find a number $\sigma > 0$ such that we have

$$\| f^s - id_D \|_{B_d} < \delta$$

for all $s \in (-\sigma, +\sigma)$. Let $m, n \in \mathbb{N}$ be large enough to have $m, n > \frac{1}{\sigma}$. Observe that (4.11) can be applied to the automorphism $f^{1/mn}$ and to the number $p = m,n$ because, for $k \leq \max\{m,n\}$, we have $k/mn < \sigma$ and, therefore,

$$\| f^{k/mn} - id_D \|_{B_d} < \delta$$

$$\| (f^{m/mn} - id_D) - m(f^{1/mn} - id_D) \|_B < \varepsilon m \| f^{1/mn} - id_D \|_B$$

or, multiplying by n,

$$\| A_n - A_{mn} \|_B \leqslant \varepsilon \| A_{mn} \|_B$$

In a similar manner we obtain

$$\| A_m - A_{mn} \|_B < \varepsilon \| A_{mn} \|_B$$

By the triangle inequality,

$$\| A_n - A_m \|_B < 2\varepsilon \| A_{mn} \|_B$$

Therefore, the proof will be complete if we establish the existence of a number $\tau > 0$ such that

$$\text{(4.15)} \qquad \sup_{|t| \leqslant \tau} \| A_t \|_B < \infty \quad \text{where} \quad A_t =: \frac{1}{t} (f^t - id_D)$$

Proof of (4.15): Let us choose τ so that we have

$$\sup_{|t| \leqslant \tau} \| f^t - id_D \|_{B_d} < \frac{d}{3}$$

Fix $t \in (-\tau/2, \tau/2)$ arbitrarily and set $n =: [\tau t^{-1}]$. Then (4.11) yields

$$\| (f^{nt} - id_D) - n(f^t - id_D) \|_B \leqslant \frac{n}{2} \| f^t - id_D \|_B ,$$

whence by the triangle inequality we obtain

$$n \| f^t - id_D \|_B < 2 \| f^{nt} - id_D \|_B$$

that is,

$$\| A_t \|_B = \frac{1}{t} \| f^t - id_D \|_B \leqslant \frac{2}{nt} \| f^{nt} - id_D \|_B < \frac{2d/3}{\tau/2}$$

since $nt \geqslant \tau/2$ and, finally

$$\| f^{nt} - id_D \|_B \leqslant \| f^{[\tau t^{-1}]t} - id_D \|_{B_d} < d/3 .$$

4.12. COROLLARY. *The vector field* A *is bounded on every ball* $B\subset\subset D$.

4.13. EXERCISE. (a) Let $U\subset\subset D$ be an open subset of U and let $(f_j)_{j\in\mathbb{N}}\subset \mathrm{Aut}\, D$ and $(t_j)_{j\in\mathbb{N}}\subset\mathbb{R}_+^*$, $t_j\to 0$, be sequences such that, for some $A\in\mathrm{Hol}(U,E)$, we have

$$\lim_{j\to\infty}\frac{1}{t_j}(f_j-\mathrm{id}_D)=A$$

with respect to the norm $\|\cdot\|_U$. Show that then, A can be holomorphically extended to D and its extension is a vector field which is complete in D.

(b) Examine the example $t\mapsto f^t$, where $f^t(\psi)=(e^{it}\psi_1,e^{2it}\psi_2,..)$ and $\psi=(\psi_1,\psi_2...)\in\ell^\infty$.

(c) Look for an unbounded complete holomorphic vector field in some bounded domain.

§3.- The Lie algebra of complete holomorphic vector fields.

4.14. DEFINITION. *The family of all holomorphic vector fields* A *on* D *that are complete in* D *will be denoted by* aut D.

4.15. LEMMA. *Let* $A\in\mathrm{aut}\, D$ *be given. If we write* $f^t=:\exp(tA)$, *then we have* $f^t\in\mathrm{Aut}\, D$ *for all* $t\in\mathbb{R}$. *Moreover, the mapping* $t\mapsto f^t$ *is T-continuous and we have the relation*

$$\mathcal{T}\lim_{t\to 0}\frac{1}{t}(f^t-\mathrm{id})=A.$$

Proof: It is well known from the elementary theory of differential equations that, given any $A\in\mathrm{Hol}(D,E)$, the set

$$U=:\{(x,t)\in D\times\mathbb{R};\ \text{there is an open interval } J \text{ with } 0,t\in J \text{ for which (4.10) has a solution on } J\}$$

is a domain in $D\times\mathbb{R}$ containing $D\times\{0\}$. The map $f(x,t)=:x^t$, unambiguously defined by (4.10) as a consequence of the local uniqueness of the solutions of differential equations, is real analytic on U and $f(\cdot,t)$ is holomorphic for each fixed t.

Furthermore, we have $f[f(x,t),s]=f(x,t+s)$ whenever (x,t), $(f(x,t),s)\in U$. Hence $f^t\in Hol(D,D)$ and $id_D=f^0=f^{-t}f^t=f^tf^{-t}$, i.e. $f^t\in Aut\ D$ for all $t\in\mathbb{R}$.

For the remainder part of the proof, let us fix any ball $B\subset\subset D$ and any $\delta>0$ such that $B_\delta\subset\subset D$ and $\|A\|_{B_\delta}<\infty$. This is possible because holomorphic maps are locally bounded. For each $x\in B$ and and $t\in\mathbb{R}$ we have

$$\|f^t(x)-x\|=\|\int_0^t(\frac{d}{ds}f^s(x))ds\|=\|\int_0^t A(f^s(x))ds\|\leqslant$$
$$\leqslant\int_0^{|t|}\|A(f^s(x))\|\,ds$$

Now we set

$$t_0=:\sup\{t\geqslant 0;\ f^s(x)\in B_\delta\ \text{ for all } s\in[-t,t] \text{ and all } x\in B\}$$

Given $t_1>t_0$ arbitrarily, there exist $x\in B$ and $s\leqslant t_1$ such that $f^s(x)\notin B_\delta$. Observe that we have $s\geqslant t_0$ necessarily. But then

$$\delta\leqslant\|f^s(x)-x\|=\|\int_0^s A(f^t(x))dt\|\leqslant\int_0^{t_1}\|A\|_{B_\delta}dt=t_1\|A\|_{B_\delta}$$

By the arbitrariness of $t_1>t_0$ we have $\delta\leqslant t_0\|A\|_{B_\delta}$. Thus $t_0>0$ and

$$\|f^t-id_D\|_B\leqslant t\|A\|_{B_\delta}$$

for all t, $|t|<t_0$. From corollary 1.7 it follows that $\mathcal{T}\lim_{t\to 0}f^t=id_D$ and therefore $\mathcal{T}\lim_{t\to s}f^t=\mathcal{T}\lim_{t\to s}f^sf^{t-s}=f^s$ for all $s\in\mathbb{R}$. But now corollary 4.10 implies that A is bounded on every ball completely interior to D, (i.e., our previous considerations are valid for each ball B and each $\delta>0$ with $B_\delta\subset\subset D$). Then

$$\frac{1}{t}[f^t(x)-x]=\frac{1}{t}\int_0^t A(f^s(x))ds=\frac{1}{t}\int_0^t[A(x)+\int_0^s\frac{d}{dr}A(f^r(x))dr]ds$$
$$=A(x)+\frac{1}{t}\int_0^t\int_0^s A^{(1}_{f^r(x)}[A(f^r(x))]dr\,ds$$

for all $x\in B$ and $t\in[-t_0,t_0]$. Therefore

$$\|\frac{1}{t}[f^t(x)-x]-A(x)\|\leqslant\frac{t}{2}\|A^{(1}\|_{B_\delta}\|A\|_{B_\delta}$$

for all $x\in B$ and t, $|t|\leq t_0$.

Let us remark that the derivative of a holomorphic map that is bounded on every interior ball, is also bounded on every interior ball. This completes the proof by the arbitrariness of $B_\delta \subset\subset D$.

With each $A\in \text{aut } D$ we associate the operator $\hat{A}: \text{Hol}(D,E)\to\text{Hol}(D,E)$ given by $\hat{A}g =: g^{(1}A$.

4.16. COROLLARY. *For any* $g\in\text{Hol}(D,E)$, *any ball* B *and any* $\delta>0$ *with* $B_\delta \subset\subset D$ *and* $\|g\|_{B_\delta}<\infty$, *there is a number* $\tau>0$ *such that we have*

$$\hat{F}^t g(x) =: g[f^t(x)] = \sum_{k=0}^{\infty} \frac{t^k}{k!} (\hat{A}\, g)(x)$$

whenever $x\in B$ *and* $|t|\leq\tau$. *The series converges uniformly on* B.

Proof: For $t>0$ we set $A =: \frac{1}{t}(f^\tau - id)$. We have $\mathcal{T}\lim_{t\to 0} A_t = A$. Therefore the statement is a consequence of (4.4).

4.17. THEOREM. *Let* A_1, $A_2 \in \text{Aut } D$ *be given. Then* A_1+A_2 *and* $A_1^{(1}A_2 - A_2^{(1}A_2$ *belong to* aut D.

Proof: Let us write $\hat{A}_g =: g\to A_g^{(1}g$ and $\hat{F}_j^t =: g\to gf^t = g[\exp(tA\ )]$ where $g\in\text{Hol}(D,E)$, $t\in\mathbb{R}$ and $j= 1,2$. Consider the net $(g_t)\subset \text{Aut } D$ defined by $g_t =: f_2^t\, f_1^t$. We show that $\mathcal{T}\lim_{t\to 0}(f_t - id_D) = A_1+A_2$. Indeed, given any ball $B\subset\subset D$, for sufficiently small values of t we have

$$g_{t|B} = \hat{F}_2^t\, \hat{F}_1^t(id_B) = [\exp t\,\hat{A}_2][\exp t\, A_1](id_B) =$$

$$= \sum_{k=0}^{\infty} \frac{t^k}{t!} \frac{d^k}{dt^k}\Big|_0 [\exp t\,\hat{A}_2][\exp t\,\hat{A}_1](id_B) =$$

$$= id_B + t(\hat{A}_2+\hat{A}_1)(id_B) + .. = id_B + (A_1+A_2) + ..$$

Therefore $\frac{1}{t}(g_t - id_D)|_B \to A_1+A_2$ in $\|\cdot\|_B$. By the arbitrariness of B, the convergence holds also in the topology $\mathcal{T}$. Then theorem 4.5 establishes that $A_1+A_2\in\text{aut } D$.

The proof of $A_1^{(1}A_2 - A_2^{(1}A_1 \in \text{aut } D$ is similar by considering the net $h_t =: f_2^t\ f_1^t\ f_2^{-t}\ f_1^{-t}$ and showing that

$$\mathcal{T}\lim_{t\to 0} \frac{1}{t^2} (h_t - id_D) = A_1^{(1} A_2 - A_2^{(1} A_1.$$

4.18. DEFINITION. *Given* $A_1, A_2 \in Hol(D,E)$, *we define*

$$[A_1, A_2] =: A_1^{(1} A_2 - A_2^{(1} A_1$$

Therefore we have

$$[A_1, A_2](x) = A_{1,x}^{(1}[A_2(x)] - A_{2,x}^{(1}[A_1(x)]$$

for $x \in D$. The operation $[,]$ is called the *Lie product* of A_1 and A_2.

For fixed $A \in Hol(D,E)$, the linear operator $[A,.]$ is called the *adjoint* of A and will be denoted by $A_{\#}$.

4.19. PROPOSITION. *For every* $A \in Hol(D,E)$, *the adjoint of* A *is derivation on* aut D, *i.e., we have*

$$A_{\#}[A_1, A_2] = [A_{\#}A_1, A_2] + [A_1, A_{\#}A_2]$$

for all $A_1, A_2 \in Hol(D,E)$.

Proof: Since the Lie product is clearly anticommutative, all we have to prove is the *Jacobi identity*

$$[A_1, [A_2, A_3]] + [A_2, [A_3, A_1]] + [A_3, [A_1, A_2]] = 0$$

for all $A_1, A_2, A_3 \in Hol(D,E)$. But

$$\begin{aligned} [A_1, [A_2, A_3]] &= A_1^{(1}[A_2, A_3] - [A_2, A_3]^{(1} A_1 = \\ &= A_1^{(1} A_2^{(1} A_3 - A_1^{(1} A_3^{(1} A_2 - (A_2^{(1} A_3 - A_3^{(1} A_2)^{(1} A_1 = \\ &= (A_1^{(1} A_2^{(1} A_3 - A_1^{(1} A_3^{(1} A_2 - A_2^{(1} A_3^{(1} A_1 + A_3^{(1} A_2^{(1} A_1) \\ &\quad - \{A_2^{(2}(A_3, A_1) - A_3^{(2}(A_2, A_1)\} \end{aligned}$$

Summing up the similar expressions for the cyclic permutations of the indexes we obtain the desired result.

#

4.20. DEFINITION. *An algebra with a product* [,] *is called a "Lie algebra" if* [,] *is anticommutative and satisfies the Jacobi identity.*

Thus we have proved

4.21. THEOREM. *The set* aut D *is a real Lie algebra with respect to the product* $[A_1,A_2]=: A_1^{(1}A_2-A_2^{(1}A_1$.

4.22. EXERCISES. (a) Show that $\frac{d^k}{dt^k}\Big|_0 (\exp tA_1)\ldots(\exp tA_n)$ belongs to the Lie subalgebra of aut D generated by $A_1,\ldots A_n$.

(b) Prove that we can write $\tau=\delta\|A\|^{-1}_{B_\delta}$ in corollary 4.13.

§4.- Some properties of commuting vector fields.

Now we turn to the investigation of holomorphic vector fields $A,B\in Hol(D,E)$ with the property $\hat{A}\hat{B}=\hat{B}\hat{A}$. In general,

$$\hat{A}\hat{B}X=\hat{A}(X^{(1}B)=(X^{(1}B)^{(1}A=2X^{(2}(B,A)+X^{(1}B^{(1}A$$

and hence

$$(\hat{A}\hat{B}-\hat{B}\hat{A})X=X^{(1}(B^{(1}A-A^{(1}B)=X^{(1}[A,B]=\widehat{[A,B]}X$$

that is,

$$\widehat{[A,B]}=\hat{B}\hat{A}-\hat{A}\hat{B}$$

Thus, $\hat{A}$ and $\hat{B}$ commute if and only if $[A,B]=0$.

Furthermore, we remark that if $X\in Hol(D,E)$ is an arbitrary vector field, then, using the argument leading to (4.7) we get

$$\|\hat{X}^n f\|_B \leqslant \left(\frac{2n}{\delta}\|X\|_B\right)^n\|f\|_{B_\delta} \tag{4.16}$$

for any open ball B with $B_\delta\subset\subset D$, any $f\in Hol(D,E)$ and $n\in\mathbb{N}$. Therefore we have

$$(\exp tX)x=\sum_{n=0}^{\infty}\frac{t^n}{n!}(\hat{X}^n id_B)x \quad x\in B_\delta$$

whenever $|t|< \frac{\delta}{2e} \|X\|_B^{-1}$ and $x\in B$ since the series converges uniformly on B by (4.16) and it clearly satisfies (4.10). For the definition of exp(tX) see Exercises 4.4.

4.23. LEMMA. *Let* $A,B\in Hol(D,E)$ *with* $[A,B]= 0$ *be given and assume that* $x\in dom(\exp A)$. *Then we have*

$$\exp(A+tB)x= (\exp tB)(\exp A)x$$

for sufficiently small values of $t\in\mathbb{R}$.

Proof: Let us write

$$f^s=: \exp sA \qquad g^t=: \exp tB \qquad h^{t,s}=: \exp(tB+A)$$

Consider the arc $S=: \{(\exp sA)x;\ \ s\in[0,1]\}$ and set

$$\delta=: \frac{1}{3}\,dist(S,\partial D) \quad , \qquad U=: S_\delta \quad , \qquad \varepsilon=: \frac{\delta}{2e}\,(\|A\|_{U_{2\delta}}+\|B\|_{U_{2\delta}})^{-1}$$

Since $\hat{A}$ and $\hat{B}$ commute, if $|s|,\ |t|<\varepsilon$ then

$$\sum_{k=0}^{\infty}\frac{s^k}{k!}\hat{A}^k\Big(\sum_{\ell=0}^{\infty}\frac{t^\ell}{\ell!}\hat{B}^\ell\Big)id_{U_{2\delta}} = \sum_{n=0}^{\infty}\frac{1}{n!}(s\hat{A}+t\hat{B})^n id_{U_{2\delta}} =$$

$$= \sum_{\ell=0}^{\infty}\frac{t^\ell}{\ell!}\hat{B}^\ell\Big(\sum_{k=0}^{\infty}\frac{s^k}{k!}\hat{A}^k\Big)id_{U_{2\delta}}$$

that is,

$$f^s g^t(y)= h^{t,s}(y)= g^t f^s(y)$$

if $|s|,\ |t|<\varepsilon$ and $y\in U_{2\delta}$.

Hence, for $s\in[0,1]$ and $|s'|,\ |t'|,\ |t|<\varepsilon$ we have

$$f^s(x)\in U \ , \qquad h^{t,s}(x)= g^t f^s(x)\in U_\delta$$

and

$$g^{t+t'}f^{s+s'}(x)= g^{t'}h^{t,s'}f^s(x)= g^{t'}f^{s'}g^t f^s(x)= h^{t',s'}g^t f^s(x)\in U_{2\delta}$$

Now, let us examine the set

$$J=: \{s\in[0,1];\ g^t f^s(x)= h^{t,s}(x) \quad \text{for} \quad |t|<\varepsilon\}$$

We knwn that $0\in J$. Let $s\in J$ be fixed and $t',s'\in(-\varepsilon,\varepsilon)$ be such that $s+s'[0,1]$. Then, from the assumption $s\in J$ we obtain

$$h^{t,s+s'}(x) = \exp[(s+s')A+tB]x=$$

$$= \exp\{ \frac{s'}{s+s'} [(s+s')A+tB]\}\exp\{ \frac{s'}{s+s'} [(s+s')A+tB]\}x=$$

$$= h^{t',s'}h^{t'',s}(x) = h^{t',s'}g^{t''}f^{s}(x) =$$

$$= g^{t'+t''}f^{s+s'}(x) = g^{t}f^{s+s'}(x)$$

where $t'=: ts'/(s+s')$ and $t''=: ts/(s+s')$, so that $s+s'\in J$. This means $J\supset [J+(-\varepsilon,\varepsilon)] \cap [0,1]$. Whence $J= [0,1]$.

#

4.24. COROLLARY. *For all* $A,B\in\mathrm{aut}D$ *with* $[A,B]= 0$ *we have*

$$\exp(A+B) = (\exp B)(\exp A).$$

Proof: For $x\in D$, we define

$$J_x =: \{t'\in\mathbb{R};\ \exp(A+t'B)x= (\exp t'B)(\exp A)x\}$$

Obviously $0\in J_x$. If $t'\in J_x$, applying lemma 4.23 to $A+t'B$ and B we obtain

$$\exp[A+(t'+t)B]x= (\exp tB)[\exp(A+t'B)]x=$$

$$= (\exp tB)(\exp t'B)(\exp A)x= [\exp(t+t')B](\exp A)x$$

for sufficiently small values of t. Thus J_x is open. However, J_x is also closed since the mappings $t'\to(\exp t'B)(\exp A)x$ and $t'\to\exp(A+t'B)x$ are continuous on $\mathbb{R}$. Therefore $J_x=\mathbb{R}$.

4.25. THEOREM. *The Lie algebra* autD *is "purely real", i.e.* $(\mathrm{aut}D) \cap i(\mathrm{aut}D) = \{0\}$.

Proof: Let $A\in(\mathrm{aut}D) \cap i(\mathrm{aut}D)$ be given. Then $\hat{A}$ commutes with $(i\hat{A})$ and, by corollary 4.23, we have

$$(\exp sA)(\exp itA) = (\exp itA)(\exp sA)$$

for all $s,t\in\mathbb{R}$. Hence

$$(\exp\xi A) = [\exp(Re\,\zeta)A]\,[\exp(Im\,\zeta)A]$$

for all $\zeta\in\mathbb{C}$. Thus, given any $x\in D$, the mapping $\phi_x: \mathbb{C}\to D$ defined by $\zeta\to(\exp\zeta A)x$ is real analytic on the whole $\mathbb{C}$. On the other hand, we have

$$\phi_x(\zeta) = \sum_{n=0}^{\infty} \frac{\zeta^n}{n!} (\hat{A}^n id_B)x$$

if $x\in B\subset\subset D$ and ζ is sufficiently small, i.e. ϕ_x is holomorphic in some neighbourhood of 0 and therefore, in view of its real analyticity, it is holomorphic on $\mathbb{C}$. Since D is bounded, ϕ_x is constant by Liouville's theorem. Thus $Ax= \frac{d}{d\zeta}\big|_0 \phi_x(\zeta) = 0$. Since x was arbitrary in D, we have A= 0.

#

§5.- The adjoint mappings.

We are going to introduce a family of mappings of fundamental importance in Lie theory.

4.26. DEFINITION. *Let* U *be any subdomain of* D *and assume that* $f: U\to D$ *and* $X: U\to E$ *are, respectively, a holomorphic mapping and a holomorphic vector field on* U. *In the sequel we shall write*

$$g_\# f =: gfg^{-1} \qquad g_\# X =: g^{(1}_{g^{-1}} X g^{-1}$$

without any danger of confusion. The maps $f\to g_\# f$ *and* $X\to g_\# X$ *(defined respectively on holomorphic local transformations and local vector fields) will be called the "adjoint maps" of* g.

By repeating the previous considerations, it is inmediate that, if $g_1: D\to D'$ and $g_2: D'\to D''$ are biholomorphic maps with $g_1(D) = D'$, $g_2(D') = D''$, then we have

$$(g_1 g_2)_\# = g_{1\#} g_{2\#} \qquad (g^{-1})_\# = (g_\#)^{-1}$$

and, by setting $X' =: g_\# X$,

$$\mathrm{dom}(\exp X') = g[\mathrm{dom}(\exp X)], \quad \exp X' = g_\#(\exp X)$$

(cf. Exercises 4.4).

4.27. EXERCISES. Let $g: D \to D'$ be a biholomorphic mapping with $g(D) = D'$. Show that:

(a) the adjoint map $g_{\#}$: $\mathrm{aut}D \to \mathrm{aut}D'$ is a surjective continuous isomorphism of the Banach-Lie algebras autD and autD'.

(b) the adjoint map $g_{\#}$: $\mathrm{Aut}D \to \mathrm{Aut}D$ is a surjective isomorphism of the topological groups AutD and AutD' endowed with their respective topologies of local uniform convergence on D and D'.

From now on we restrict ourselves to the case in which $D = D'$.

4.28. THEOREM. *Let us suppose that* $(g_j)_{j \in J}$ *and* $(t_j)_{j \in J}$ *are, respectively, nets in* AutD *and* $\mathbb{R}$ *with*

$$\lim_{j\to\infty} t_j = 0 \qquad \mathcal{T}\lim_{j\to\infty} (g_j - id_D) = A \in \mathrm{aut}D$$

Consider any holomorphic local transformation $f: U \to D$ *and a vector field* $X: U \to E$. *Then, for any ball* B *such that* $B_\delta \subset\subset D$, *we have*

$$\frac{1}{t_j} (g_{j\#} f - f)|_B \to (Af - f^{(1}A)|_B$$

and

$$\frac{1}{t_j} (g_{j\#} X - X)|_B \to (A^{(1}X - X^{(1}A)|_B$$

uniformly on B.

Proof: Given B with $B_\delta \subset\subset D$, if $\| t_j A \|_{B_\delta} < \frac{2\delta}{e}$ we have $g_j^{-1} B \subset U$. Therefore

$$\frac{1}{t_j} (g_{j\#} f - f)|_B = \frac{1}{t_j} (g_j f g_j^{-1} - f)|_B = \frac{1}{t_j} (g_j f g_j^{-1} - f g_j^{-1})|_B + \frac{1}{t_j} (f g_j^{-1} - f)|_B =$$

$$= \frac{1}{t_j} (g_j - id_D) f g_j^{-1}|_B + \frac{1}{t_j} (f g_j^{-1} - f)|_B \to (Af - f^{(1}A)|_B$$

uniformly on B. Similarly, we have

$$\frac{1}{t_j}(g_j\ X-X)|_B = \frac{1}{t_j}(g^{-1}_{g_j^{-1}} X g_j^{-1} - X)|_B =$$

$$= \frac{1}{t}(g_j^{(1}{}_{g_j^{-1}} X g_j^{-1} - X g_j^{-1})|_B + \frac{1}{t_j}(X g_j^{-1} - X)|_B =$$

$$= \frac{1}{t_j}[(g_j - id_D)^{(1}_{g_j^{-1}}] X g_j^{-1}|_B + \frac{1}{t_j}(X g_j^{-1} - X)|_B \to$$

$$\to A^{(1}X - X^{(1}A|_B = [A,X]|_B .$$

uniformly on B, where $[A,X] =: [A_{|U}, X_{|U}]$.

<u>4.29. EXERCISE</u>. Give a detailed proof of these formulas Hint: the essential tools have been developed in the proof of proposition 4.1.

In particular, if $g^t = \exp tA$ and t is so small that $g^{-t}B \subset U$, we have

$$\frac{1}{h}(g_{\#}^{t+h}X - g_{\#}^{t}X)|_B = \frac{1}{h}[g_{\#}^{h}(g_{\#}^{t}X) - g_{\#}^{t}X]|_B \to [A, g_{\#}^{t}X]|_B$$

with respect to the norm $\|\cdot\|_B$. Therefore $Y(t) =: g_{\#}^{t}X_{|B}$ satisfies the differential equations

$$\frac{d}{dt}Y(t) = [A, Y(t)]|_B = (A_{|B})_{\#}Y(t)$$

where the derivative is taken in the sense of the norm $\|\cdot\|_B$. Hence we have

$$Y(t) = \sum_{n=0}^{\infty} \frac{t^n}{n!}(A_{|B})_{\#}^{n}(X_{|B})$$

for sufficiently small values of t, because

$$\| A_{\#}^{n}X \|_B \leq \left(\frac{2n}{\delta}\|A\|_{B_\delta}\right)^n \|X\|_{B_\delta}$$

for $n \in \mathbb{N}$. Indeed, if $B'_{\delta'} \subset\subset D$ and $Z \in Hol(D,E)$ we have

$$\| A_{\#} Z \|_{B'} = \| A^{(1} Z - Z^{(1} A \|_{B'} \leqslant$$

$$\leqslant \| A^{(1} \|_{B'} \| Z \|_{B'} + \| Z^{(1} \|_{B'} \| A \|_{B'} \leqslant$$

$$\leqslant \frac{1}{\delta'} \| A \|_{B'_{\delta'}} \| Z \|_{B'} + \frac{1}{\delta'} \| Z \|_{B'_{\delta'}} \| A \|_{B'} \leqslant$$

$$\leqslant \frac{2}{\delta} \| A \|_{B'_{\delta'}} \| Z \|_{B'_{\delta'}}$$

and so

$$\| A_{\#}^{n} \|_{B} \leqslant \| A_{\#} (A_{\#}^{n-1} X) \|_{B} \leqslant \frac{2}{\delta/n} \| A \|_{B_{\delta/n}} \| A_{\#}^{n-1} \|_{B_{\delta/n}} \leqslant$$

$$\leqslant \ldots \leqslant \frac{2}{\delta/n} \| A \|_{B_{\delta/n}} \frac{2}{\delta/n} \| A \|_{2\delta/n} \cdots \frac{2}{\delta/n} \| A \|_{n\delta/n} \| X \|_{n\delta/n}$$

That is, roughly speaking we have

$$(\exp tA)_{\#} = \exp(tA_{\#})$$

locally. We shall see in the next chapter in which sense the series $\sum_{n=0}^{\infty} \frac{t^n}{n!} A_{\#}^{n} X$ converges globally if $X \in \mathrm{aut} D$.

CHAPTER 5

THE NATURAL TOPOLOGY ON THE LIE ALGEBRA OF COMPLETE VECTOR FIELDS

§1. Cartan's uniqueness theorem for autD.

5.1. THEOREM. *Let* $A\in\operatorname{aut}D$ *be given and suppose that* $A_a^{(0}=A_a^{(1}=0$ *for some* $a\in D$. *Then* $A=0$.

Proof: Set $f^t(x)=:(\exp tA)x$. Observe that $f^t(a)=a$ for all $t\in\mathbb{R}$, since

$$\frac{d}{dt}f^t(a)=A[f^t(a)]=0$$

is satisfied in this case. Let us compute $(f^t)_a^{(1}$. By the analyticity of the mapping $(t,y)\to f^t(y)$ we have

$$\frac{d}{dt}(f^t)_x^{(1}h=\frac{\partial}{\partial t_1}\frac{\partial}{\partial t_2}\Big|_0 f^{t+t_1}(x+t_2h)=$$

$$=\frac{\partial}{\partial t_2}\frac{\partial}{\partial t_1}\Big|_0 f^{t+t_1}(x+t_2h)=\frac{\partial}{\partial t_2}\Big|_0 A[f^t(x+t_2h)]=$$

$$=A_{f^t(x)}^{(1}[(f^t)_x^{(1}h]$$

for all $x\in D$. Thus we have

$$\frac{d}{dt}(f^t)_x^{(1}=A_{f^t(x)}^{(1}f_x^{t(1}\tag{5.1}$$

for all $t\in\mathbb{R}$.

In particular

$$\frac{d}{dt}(f^t)_a^{(1}=A_a^{(1}f_a^{t(1}=0$$

whence $(f^t)_a^{(1} = (f^t)_a^{(0} = \mathrm{id}$ for all $t\in\mathbb{R}$. Then, by Cartan's uniqueness theorem, we have $f^t = \mathrm{id}_D$ for all $t\in\mathbb{R}$, whence $A = 0$.

#

5.2. COROLLARY. *Given any* $a\in D$, *the mapping* $\mathrm{aut}D\to E\times L(E|E)$ *defined by* $A\to(A_a^{(0}, A_a^{(1})$ *is injective and continuous.*

Proof: The continuity is clear. If $A_{1,a}^{(s} = A_{2,a}^{(s}$ for $s = 0,1$, then $A_1 - A_1 \in \mathrm{aut}D$ and $(A_1 - A_2)_a^{(s} = 0$ for $s = 0,1$.

#

§2.- Some majorizations on autD.

The previous corollary naturally raises this question: Is the inverse mapping $(A_a^{(0}, A_a^{(1}) \to A$ continuous?. By its linearity, its continuity means some kind of "lipschitzianity". To obtain a result in this direction, we go back to AutD.

5.3. THEOREM. *Given a ball* $B \subset\subset D$ *centered at* $a\in D$, *there exists a constant* K *such that we have*

$$\| f - \mathrm{id}_D \|_B \leq K \sum_{s=0}^{1} \| f_a^{(s} - \mathrm{id}_{D\,a}^{(s} \|$$

for all $f\in\mathrm{Aut}D$.

Proof: We proceed by contradiction. Let $(f_j)_{j\in\mathbb{N}} \subset \mathrm{Aut}D$ be a sequence with $f_j \neq \mathrm{id}_D$ for $j\in\mathbb{N}$ such that

$$\| f_j - \mathrm{id}_D \|_B = \mu_j \sum_{s=0}^{1} \| (f_j - \mathrm{id}_D)_a^{(s} \| \tag{5.2}$$

where $\mu_j \to \infty$.

Since $\| f_j - \mathrm{id}_D \|_B \leq \mathrm{diam}D$, we have $\sum_{s=0}^{1} \| (f_j - \mathrm{id}_D)_a^{(s} \| \to 0$, i.e., by theorem 2.7, $\mathcal{T}\lim_{j\to\infty} f_j = \mathrm{id}_D$. Let us fix $d>0$ such that $B_\delta \subset\subset D$ and define $p_j =: 1+\max\{p\in\mathbb{N};\ \| f_j^k - \mathrm{id}_D \|_{B_d} < d/4 \text{ for } 0\leq k\leq p\}$.

From lemma 4.8 we know that

$$p_j \leq 1 + \frac{d}{2\| f_j - \mathrm{id}_D \|_B}$$

for all $j\in\mathbb{N}$. For the sake of simplicity we may assume that we have

$$p_j \leqslant \frac{d}{\| f_j - id_D \|_B} \tag{5.3}$$

for all $j \in \mathbb{N}$. By definition we have $f_j^{p_j} \not\to id_D$ with respect to the norm $\| \cdot \|_{B_d}$. However, we shall show that

$$(f_j^{p_j})_a^{(0} \to a \quad \text{and} \quad (f_j^{p_j})_a^{(1} \to id$$

which is a contradiction by theorem 2.7. An application of corollary 4.6 yields

$$\| \, |f_j^p(a)-a| - p|f_j(a)-a| \, \| \leqslant$$

$$\leqslant \frac{1}{d-d/4} \| f_j(a)-a \| \sum_{k=1}^{p-1} \| f^k - id \|_{B_d(a)} \leqslant$$

$$\leqslant \frac{2}{d} p \frac{d}{4} \| f_j(a)-a \| = \frac{p}{2} \| f_j(a)-a \|$$

for all $j \in \mathbb{N}$ and $p \leqslant p_j$. Therefore, by (5.3),

$$\| f_j^p(a)-a \| \leqslant \frac{3}{2} p_j \| f_j(a)-a \| \leqslant 2d \frac{\| f_j(a)-a \|}{\| f_j - id_D \|_D} = \tag{5.4}$$

$$= \frac{2d \| f_j(a)-a \|}{\mu_j \sum_{s=0}^{1} \| (f_j - id_D)_a^{(s} \|} \leqslant \frac{2d}{\mu_j} \to 0$$

for all $p \leqslant p_j$.

Now, let us consider the sequence $(f_j^{p_j})_a^{(1}$. We have

$$(f_j^{p_j})_a^{(1} = (f_j)^{(1}_{f_j^{p_j-1}(a)} \cdots (f_j)_a^{(1} = \prod_{k=1}^{p_j} (id + H_{jk})$$

where

$$H_{jk} =: (f_j)^{(1}_{f_j^{p_j-1}(a)} - id$$

for $k = 1,2,\ldots,p_j$ and $j\in\mathbb{N}$. Therefore

$$\|(f_j^{p_j})_a^{(1}-id\| = \|\prod_{k=1}^{p_j}(id+H_{jk})\| = \|\sum_{J\subset\{1,..,p_j\}}\prod_{k\in J}H_{jk}\| \leq$$

$$\leq \sum_{J\subset\{1,..,p_j\}}\prod_{k\in J}\|H_{jk}\| = \prod_{k=1}^{p_j}(1+\|H_{jk}\|)-1$$

We must prove that the right-hand side tends to 0. To do this, let us denote by r the radius of B and choose $j_0\in\mathbb{N}$ such that $2d/\mu_d<r/2$ for all $j\geq j_0$. Then, by (5.4) we have

$$f_j^{k-1}(a)\in B' =: B_{r/2}(a)$$

for all $j\geq j_0$ and $k = 1,\ldots,p_j$. Therefore, by proposition 1.4 and (5.2)

$$\|H_{jk}\| = \|(f_j)^{(1}_{f_j^{k-1}(a)}-id\| \leq$$

$$\leq 2\left(\frac{2e}{r}\|f_j-id_D\|_B\right)^2\|f_j^{k-1}(a)-a\| + \frac{1}{\mu_j}\|f_j-id_D\|_B$$

whenever $j\geq j_0$ and $1\leq k\leq p_j$. Taking into account (5.4) and the fact that $\|f_j-id_D\|_B\to 0$, we can write

$$\|H_{jk}\| \leq \gamma\frac{1}{\mu_j}\|f_j-id_D\|_B$$

for some constant γ (independent of j,k) and all $j\geq j_0$, $k = 1,..,p_j$. Hence,

$$(5.5)\qquad \|(f^{p_j})_a^{(1}-id\| \leq \left(1+\gamma\frac{1}{\mu_j}\|f_j-id_D)\|_B\right)^{p_j}$$

for $j\geq j_0$. It is well known from elementary analysis that $(1+\alpha_j)^{\beta_j}\to 1$ whenever $\alpha_j\to 0$ and $\alpha_j\beta_j\to 0$. But, by (5.3),

$$p_j\frac{1}{\mu_j}\|f_j-id_D\|_B \leq \frac{d}{\mu_j}\to 0.$$

#

5.4. COROLLARY. *For every pair of balls* B *and* B' *with* $B, B' \subset\subset D$ *there exists a constant* K' *such that we have*

$$\| f-g \|_B \leqslant K' \sum_{s=0}^{1} \| f_a^{(s} - g_a^{(s} \|$$

for all $f, g \in \mathrm{Aut}D$ *satisfying* $f(B) \subset B'$ *and* $g(B) \subset B'$.

Proof: Given $f, g \in \mathrm{Aut}D$ with $f(B) \subset B'$ and $g(B) \subset B'$, by theorem 5.3 and remark 2.4 we have

$$\| f-g \|_B \leqslant K_1 \| g^{-1}f - id_D \| \leqslant K_2 \sum_{s=0}^{1} \| (g^{-1}f - id_D)_a^{(s} \| \leqslant$$

$$\leqslant K_2 \| g^{-1}f(a) - g^{-1}g(a) \| + K_2 \| (g^{-1})_{f(a)}^{(1} f_a^{(1} - (g^{-1})_{g(a)}^{(1} g_a^{(1} \| \leqslant$$

$$\leqslant K_2 \| g^{(1} \|_{B'} \| f(a) - g(a) \| + K_2 \| (g^{-1})_{f(a)}^{(1} (f_a^{(1} - g_a^{(1}) \| +$$

$$+ K_2 \| [(g^{-1})_{f(a)}^{(1} - (g^{-1})_{g(a)}^{(1}] \| \leqslant K_2 \{ \| g^{(1} \|_{B'} \| f(a) - g(a) \| +$$

$$+ \| (g^{-1})^{(1} \|_{B'} \| f_a^{(1} - g_a^{(1} \| + \| (g^{-1})^{(2} \|_{B'} \| g^{(1} \|_B \| f(a) - g(a) \| \}$$

where, by proposition 1.4, the right-hand side is dominated by

$$K_3 \sum_{s=0}^{1} \| f_a^{(s} - g_a^{(s} \|$$

for some constants K_1, K_2, K_3 depending only on B and B'.

#

5.5. EXERCISE. Look for counterexamples to show that the constant K' in corollary 5.4 must actually depend on B and B'.

§3.- The natural topology on autD.

From the strong statement of theorem 5.3 it is already easy to deduce a result concerning autD:

5.6. THEOREM. *Given any ball* $B \subset\subset D$ *centered at* $a \in D$, *there exists a constant* K_B *such that we have*

(5.6) $$\|A\|_B \leq K_B \sum_{s=0}^{1} \|A_a^{(s}\|$$

for all $A \in autD$. *If* $B' \subset\subset D$ *is another ball centered at* $a' \in D$, *we have*

$$\|\cdot\|_B \sim \sum_{s=0}^{1} \|\cdot_a^{(s}\| \sim \|\cdot\|_{B'}$$

on autD.

Proof: Theorem 5.3 furnishes a constant K_B such that

$$\|f-id_D\|_B \leq K_B \sum_{s=0}^{1} \|f_a^{(s}-id_a^{(s}\|$$

for all $f \in AutD$. Hence, given $A \in autD$ and writing $f^t =: \exp tA$, $t \in \mathbb{R}$, we have

$$\|\frac{1}{t}(f^t-id_D)\|_B \leq K_B \sum_{s=0}^{1} \|\frac{1}{t}(f^t-id_D)_a^{(s}\|$$

for all $t \in \mathbb{R}$. But $T\lim_{t\to 0} \frac{1}{t}(f^t-id_D) = A$ whence $\frac{1}{t}(f^t-id_D)$ tends to A also in the norms $\|\cdot\|_B$ and $\sum_{s=0}^{1} \|\cdot_a^{(s}\|$. This proves (5.6). To prove that

$$\|\cdot\|_B \sim \sum_{s=0}^{1} \|\cdot_a^{(s}\| \sim \|\cdot\|_{B'}$$

we need only to copy the proof of theorem 1.6. #

5.7. REMARK. The equivalence of two norms on a vector space implies the existence of $\mu_1, \mu_2 > 0$ such that we have $\mu_1 \|x\|_1 \leq \|x\|_2 \leq \mu_2 \|x\|_1$ for all x.

§4.- autD as a Banach space.

After the previous theorem, the next question is at hand: Is autD endowed with any of the norms $\|\cdot\|_B$ or $\sum_{s=0}^{1} \|\cdot_a^{(s}\|$ a Banach space?. That is, given a sequence $(A_j)_{j\in\mathbb{N}} \subset autD$ with $A_{ja}^{(s} \to L^{(s}$, s= 0,1, does there exist $A \in autD$ such that we have $A_a^{(s} = L^{(s}$ for s= 0,1? We can prove a much stronger result that has crucial importance in the theory of symmetric domains.

5.8. THEOREM. *Let the nets* $(f_j)_{j\in J} \subset AutD$ *and* $(t_j)_{j\in J} \subseteq \mathbb{R}_+^*$ *be given and assume that we can find some* $a \in D$, *some* $L^{(0} \in E$ *and some* $L^{(1} \in \mathcal{L}(E|E)$ *such that the net*

$$A_j =: \frac{1}{t_j}(f_j - id_D)$$

satisfies $A^{(s}_{j,a} \to L^{(s}$ *for* s= 0,1. *Then, there exists* $A \in autD$ *such that we have* $T\lim_{j\in J} A_j = A$.

Proof: Let us fix a ball $B \subset\subset D$ centered at $a \in D$ and choose $d>0$ such that $B_{2d} \subset\subset D$. We may assume

$$\sup_{j\in J} \sum_{s=0}^{1} \| A^{(s}_{ja} \| < \infty$$

Then, by theorem 5.3 there exists $M>0$ such that

$$\| f_j - id_D \|_{B_{2d}} \leqslant Mt_j$$

for all $j \in J$. For $\delta < d/4$ we define

$$p_j(\delta) =: 1 + \max\{p \in \mathbb{N};\ \| f_j^p - id_D \|_{B_d} < \delta\}$$

From lemma 4.8 we see that

$$p(\delta) \geqslant \frac{\delta}{\| f_j - id \|_{B_{2d}}} \geqslant \frac{\delta}{Mt_j}$$

We set $t =: \delta/M$ and $n_j =: [tt_j^{-1}]$ where we keep t and δ fixed for the moment. Note that $n_j \leqslant p_j(\delta)$ for all $j \in J$. Therefore, by corollary 4.6 we have

$$(5.7)\quad \| (f_j^{n_j} - id_D) - n_j(f_j - id_D) \|_B \leqslant \frac{n_j\delta}{d-\delta} \| f_j - id_D \|_B \leqslant$$

$$\leqslant \frac{n_j\delta}{d/2} Mt_j \leqslant \frac{M}{d/2}\delta t$$

Denote by r the radius of B. From (5.7) using the Cauchy estimates we obtain

$$\| (f_j^{n_j} - id_D)^{(s}_a - n_j t_j A^{(s}_{j,a} \| \leqslant \max(1,r) \| (f^{n_j} - id_D) - n_j t_j A_j \|_B \leqslant M_2 \delta t$$

for s= 0,1. Now, the triangle inequality applied to the segments suggested by the below diagram

$$f_j^{n_j}-id_D \qquad\qquad f_k^{n_k}-id_D$$

$$n_j t_j A_j = n_j(f_j-id_D) \qquad\qquad n_k t_k A_k = n_k(f_k-id_D)$$

implies

$$\| (f_j^{n_j}-f_k^{n_k})_a^{(s} \| \leqslant 2M_2 \delta t + \| n_j t_j A_{j,a}^{(s} - n_k t_k A_{k,a}^{(s} \|$$

Observe that

$$\lim_{j,k} \| n_j t_j A_{j,a}^{(s} - n_k t_k A_{k,a}^{(s} \| = \| tA_a^{(s} - tA_a^{(s} \| = 0$$

for s= 0,1. Hence, there exists and index $j_0(\delta)$ such that

$$\|(f_j^{n_j}-f_k^{n_k})_a^{(s}\| \leqslant M_3 \delta t$$

for all j, $k \geqslant j_0(\delta)$ and s= 0,1. Since $n_j \leqslant p_j(\delta)$, from the definition of $p_j(\delta)$ we see that $f_j^{n_j}(B) \subset B_d$ for all $j \in J$. Thus we may apply corollary 5.4 to $B' =: B_d$ and $f =: f_j^{n_j}$, $g = f_k^{n_k}$. Hence

$$\| f_j^{n_j} - f_k^{n_k} \|_B \leqslant M_4 \delta t$$

ro all j, $k \geqslant j_0(\delta)$, where M_4 is independent of j, k, δ and t. Applying this result to (5.7), the triangle inequality yields

$$f_j^{n_j}-id_D \qquad f_k^{n_k}-id_D$$

$$n_j t_j A_j \qquad n_k t_k A_k$$

$$\| n_j t_j A_j - n_k t_k A_k \|_B \leqslant \frac{2M}{d/2}\delta t + M_4 \delta t$$

that is,

$$(5.9) \qquad \| [tt_j^{-1}]^{t_j t^{-1}} A_j - [tt_k^{-1}]^{t_k t^{-1}} A_k \|_B \leqslant M_4 \delta$$

for some $M_4>0$, all j, $k\geq j_0(\delta)$ and all $t\in(0,\delta/M)$. From (5.9) it readily follows (by taking the superior limit in j,k with fixed δ,t) that

$$\limsup_{j,k} \| A_j - A_k \|_B \leq M_4 \delta$$

for all $\delta>0$, i.e., $(A_j)_{j\in J}$ is a Cauchy net with regard to the norm $\|\cdot\|_B$. Consequently, we have $(A_j)_x^{(s} \to L^{(s,x)}$, $s=0,1$, for some $L^{(0,x)}\in E$ and $L^{(1,x)}\in \mathcal{L}(E|E)$ whenever $x\in B$. But then $(A_j)_{j\in J}$ is a Cauchy net with regard to the norm $\|\cdot\|_{B'}$ whenever $B'\subset\subset D$ is a ball centered at a point $x\in B$. By repeating the argument, from the connectedness of D we obtain that $(A_j)_{j\in J}$ is a Cauchy net with regard to the norm $\|\cdot\|_{B''}$ for any ball $B''\subset\subset D$, i.e. $(A_j)_{j\in J}$ is a $\mathcal{T}$-Cauchy net. Thus we have $\mathcal{T}\lim_{j\in J} A_j = A$ for some $A\in Hol(D,E)$. But then, theorem 4.3 establishes that $A\in autD$.

5.9. COROLLARY. *Let* $B\subset\subset D$ *any ball centered at* $a\in D$. *Then* autD *is Banach space with regard to the norms* $\|\cdot\|_B$ *and* $\sum_{s=0}^{1} \|\cdot_a^{(s}\|$.

Proof: Let us suppose that $(A_j)_{j\in\mathbb{N}}$ is a Cauchy sequence in the norm $\sum_{s=0}^{1}\|\cdot_a^{(s}\|$. Then $A_{j,a}^{(s}\to L^{(s}$ for $s=0,1$.

Choose a sequence $(\varepsilon_j)_{j\in\mathbb{N}}$ of positive numbers with $\varepsilon_j\to 0$. Since we have

$$\mathcal{T}\lim_{t\to 0} \frac{1}{t}(\exp tA_j - id_D) = A_j$$

we can pick $t_j>0$ such that $t_j<\varepsilon_j$ and

$$\sum_{s=0}^{1} \left\| \frac{1}{t_j}(f_j - id_D)_a^{(s} \right\| < \varepsilon_j$$

where $f_j =: \exp t_j A_j$. Obviously we have

$$\frac{1}{t_j}(f_j - id_D)_a^{(s} \to L^{(s}$$

for $s=0,1$. Now theorem 5.8 ensures that $L^{(s} = A_a^{(s}$ for some

$A\in autD$.

#

5.10. EXERCISE. Prove corollary 5.9 directly by using the local uniform continuity of the solutions of ordinary differential equations with regard to the initial values.

§.- autD as a Banach-Lie algebra.

Let us fix any ball $B\subset\subset D$ and $\delta>0$ such that $B_\delta\subset\subset D$, and endow autD with the norm $\|\cdot\|_B$. We already know that $(autD, \|\cdot\|_B)$ is a Banach space. Let us now consider its Lie-algebra structure.

5.11. LEMMA. *For all* $A\in autD$, *the mapping* $A_\#: X\to[A,X]$ *is a bounded linear operator on autD.*

Proof: The linearity of $A_\#$ is obvious. On the other hand, by the Cauchy estimates and the fact $\|\cdot\|_B\sim\|\cdot\|_{B'}$ we have

$$\|[A,X]\| = \|A^{(1}X-X^{(1}A\|\leqslant\|A^{(1}\|_B\|X\|_B+\|X^{(1}\|_B\|A\|_B\leqslant$$

$$\leqslant\frac{1}{\delta}\|A\|_{B_\delta}\|X\|+\frac{1}{\delta}\|X\|_{B_\delta}\|A\|\leqslant$$

$$\leqslant\frac{M}{\delta}\|A\|_B\|X\|+\frac{M}{\delta}\|X\|_B\|A\|=\frac{2M}{\delta}\|A\|\ \|X\|$$

for all $X\in autD$ and some M (independent of X).

5.12. COROLLARY. *The mapping* $\#: A\to A_\#$ *is a continuous linear operator on* autD.

Proof: We have $\|A_\#\|\leqslant M'\|A\|$ for all $A\in autD$ and some $M'>0$.

5.13. PROPOSITION. *We have*

$$\exp(A_\#) = (\exp A)_\#$$

for all $A\in autD$.

Proof: Let $X\in autD$ be arbitrarily fixed. By lemma 5.11 we have that

$$[\exp(tA_{\#})]X =: \sum_{k=0}^{\infty} \frac{t^k}{k!} A_{\#}^k X$$

is a well-defined element $\tilde{Y}(t)$ of autD. Moreover,

$$\frac{d}{dt}\tilde{Y}(t) = \lim_{h\to 0} \frac{1}{h}\left[\tilde{Y}(t+h)-\tilde{Y}(t)\right] = A_{\#}\tilde{Y}(t)$$

for all $t\in\mathbb{R}$ and $\tilde{Y}(0) = X$. But the norm convergence of $\frac{1}{h}\left[\tilde{Y}(t+h)-\tilde{Y}(t)\right]$ means its $\mathcal{T}$-convergence in view of corollary 5.12. Thus the mapping $t\to\tilde{Y}(t)$ satisfies the differential equation

$$\frac{dy}{dt} = A_{\#}[y(t)], \qquad y(0) = X \tag{5.10}$$

in the Banach space (autD, $\mathcal{T}$). But we have seen in §4 Chapter IV, that the mapping $Y(t) = (\exp tA)_{\#}X$, $t\in\mathbb{R}$, satisfies this equation, too, whence the result follows.

#

5.14. LEMMA. *Let* ϕ *be a continuous automorphism of the Banach Lie algebra* autD. *Then we have*

$$(\exp\phi A)_{\#}\phi X = \phi[\exp(A_{\#})]X$$

for all A, X∈autD.

Proof: Since ϕ is an automorphism of the Lie algebra autD, we have

$$\phi(A_{\#}X) = \phi[A,X] = [\phi A,\phi X] = (\phi A)_{\#}\phi X$$

and, by reiterating the argument we obtain

$$\phi[(A_{\#})^n X] = (\phi A_{\#})^n \phi X$$

for $n\in\mathbb{N}$. As ϕ is a continuous linear operator on autD, by proposition 5.13,

$$\phi[(\exp A)_{\#}X] = \phi[\exp(A_{\#})X] = \phi\left[\sum_{n=0}^{\infty} \frac{1}{n!}(A_{\#})^n X\right] =$$

$$= \sum_{n=0}^{\infty} \frac{1}{n!}\phi[(A_{\#})^n X] = \sum_{n=0}^{\infty} (\phi A_{\#})^n \phi X = (\exp \phi A_{\#})\phi X.$$

#

CHAPTER 6

THE BANACH LIE GROUP STRUCTURE OF THE SET OF AUTOMORPHISMS

We have seen that AutD is a topological group when endowed with the topology $\mathcal{T}$ of local uniform convergence. Now we are going to construct another topology $\mathcal{T}_a$ on AutD such that (AutD, $\mathcal{T}_a$) carries the structure of a real Banach-Lie group which acts analytically on D. First we introduce some preparatory material.

§1.- The concept of a Banach manifold.

Let M and E be respectively a Hausdorff space and a Banach space over any of the fields $\mathbb{R}$ or $\mathbb{C}$ which we indistinctly represent by $\mathbb{K}$.

6.1. DEFINITION. *A "chart" of* M *over* E *is a pair* (U,u) *where* U *is an open subset of* M *and* u *is a homeomorphism of* U *onto an open subset of* E.

An "analytic structure" on M *is a collection of charts* $(U_\alpha, u_\alpha)_{\alpha\in I}$ *of* M *over* E *such that the following conditions are satisfied:*

M_1: *The famuly* $(U_\alpha)_{\alpha\in I}$ *is an open cover of* M.

M_2: *For each pair* $\alpha,\beta\in I$, *the maping* $\mu_\beta\circ\mu_\alpha^{-1}$: $\mu_\alpha(U_\alpha\cap U_\beta)\to\mu_\beta(U_\alpha\cap U_\beta)$ *is analytic.*

M_3: *The collection* $(U_\alpha,u_\alpha)_{\alpha\in I}$ *is a maximal family of charts on* M *for which conditions* M_1 *and* M_2 *hold.*

A "Banach manifold" is a pair (M,*A*) *where* M *is a Hausdorff space and* *A* *is an analytic structure on* M *over some Banach space* E.

If there is no danger of confusion, we shall refer to the Banach manifold M without any reference to its analytic structure *A*.

According as the field $\mathbb{K}$ is $\mathbb{R}$ or $\mathbb{C}$ we say that M is a *real* or a *complex manifold.*

6.2. REMARK. Condition M_3 will often be cumbersome to check in specific instances. In fact, if conditions M_1 and M_2 are satisfied, the family $(U_\alpha,u_\alpha)_{\alpha\in I}$ can be extended in a unique manner to a larger family of charts for which condition M_3 is satisfied, too. Thus, M_3 is not essential in the definition of a Banach manifold.

6.3. EXEMPLES. Let U be a non void open subset of a Banach space E. The pair (U,id_U) is a chart of U over E and defines analytic structure on U. The manifold so constructed is called the *canonical manifold* on U.

Let M and N be Banach manifolds over the Banach spaces E and F respectively. If (U,u) and (V,v) are charts of M and N, then the pair $(U\times V, u\times v)$, where $u\times v\colon (x,y)\to (u(x),v(y))$, is a chart of $M\times N$ over $E\times F$. The family of the pairs so constructed is an analytic structure and the corresponding manifold is called the *product* of M and N.

Let M be a Banach manifold over a complex Banach space E. Then E can be considered as a real Banach space, too, which we denote by $E_{\mathbb{R}}$. Any chart (U,u) of M over E is a chart over $E_{\mathbb{R}}$, and the family of these charts defines a real analytic structure on M. The manifold so constructed is called the *underlying real manifold* of M.

6.4. DEFINITION. *Let a Banach manifold* M *and a point* $x\in M$ *be given, and consider the set of the pairs* $[(U,u),h]$ *where* (U,u) *is a chart of* M *at* x *and* $h\in E$. *We say that* $[(U,u),h_1]$ *and* $[(V,v),h_2]$ *are "equivalent" if we have*

$$(v\circ u^{-1})^{(1}_{u(x)}\cdot h_1 = h_2$$

We write T_xM for the quotient set. The equivalence class of the element $[(U,u),h]$, which is denoted by $h\,\frac{\partial}{\partial u}\big|_x$, is called a *tangent vector* to M at x.

Let us fix any chart (U,u) of M at $x\in M$. The mapping $E\to T_xM$ given by $h\to h\frac{\partial}{\partial u}\Big|_x$ is a bijection by means of which we can transfer the Banach space structure of E to T_xM. We say that T_xM endowed with this Banach space structure is the *tangent space* to M at x.

6.5. DEFINITION. *Let a Banach manifold* M *and a Banach space* F *be given. We say that a mapping* $f\colon M\to F$ *is "analytic at a point* $x\in M$*" if there is a chart* (U,u) *of* M *at* x *such that* $f\circ u^{-1}\colon u(U)\to F$ *is analytic. We say that* f *is "analytic on* M*" if it is analytic at every point* $x\in M$ *and we call* $f\circ u^{-1}$ *a "local expression" of* f *at* x.

Let $f, g\colon M\to F$ be analytic mappings at a point $x\in M$, and denote by $f\circ u^{-1}\colon u(U)\to F$, $g\circ v^{-1}\colon v(V)\to F$ their local representations in the charts (U,u) and (V,v), respectively. We say that f and g are *equivalent at* x if there is a neighbourhood $W\subset U\cap V$ of x such that $f\circ u^{-1}= g\circ v^{-1}$ on W. We denote by θ_x^F the quotient set and each equivalence class is called an *analytic germ* at x. θ_x^F is endowed with a vector space structure in an obvious manner.

Now, tangent vectors to M at x can be interpreted as differential operators acting on analytic germs at x in the following manner: for $f\in\theta_x^F$ and $h\frac{\partial}{\partial u}\Big|_x\in T_xM$ we set

$$\frac{\partial f}{\partial u}(x).h =: (f\circ u^{-1})^{(1}_{u(x)}h$$

6.6. DEFINITION. *Let* M *and* N *be Banach manifolds over the Banach spaces* E *and* F, *respectively. We say that a continuous mapping* $f\colon M\to N$ *is a "morphism" of Banach manifolds if, for each point* $x\in M$, *there are charts* (U,u) *of* M *at* x *and* (V,v) *of* N *at* $y= f(x)$ *such that* $v\circ f\circ u^{-1}\colon u(U)\to v(V)$ *is analytic.*

Suppose that $f\colon M\to N$ is a morphism of Banach manifolds. Then $(v\circ f\circ u^{-1})^{(1}_{u(x)}$ is an element of $\mathcal{L}(E,F)$ and we can define a continuous linear mapping $df(x)\colon T_xM\to T_{f(x)}N$ by setting

$$(6.1)\qquad df(x)\colon h\frac{\partial}{\partial u}\Big|_x \to (v\circ f\circ u^{-1})^{(1}_{u(x)}.h\frac{\partial}{\partial v}\Big|_{f(x)}$$

for $h \in E$. It is easy to check that $df(x)$ does not depend on the charts (U,u) and (V,v) we have chosen. We say that $df(x)$ is the *derivative* of f at x and that (6.1) is its *local expression* with respect to the charts (U,u) and (V,v).

Let M be a Banach manifold over E and let U be an open subset of M. We set

$$TU =: \{T_x M;\ x \in U\}$$

If (U,u) is a chart of M, we define a mapping $T_u: TU \to u(U) \times E$ by means of

$$T_u: h \frac{\partial}{\partial u}\Big|_x \to (u(x),h).$$

Then, we have

6.7. PROPOSITION. *There exists a unique topology on* TM *such that the following conditions are satisfied:*

(a) *For every open subset* U *of* M, TU *in an open subset of* TM.

(b) *For every chart* (U,u) *of* M, *the mapping* $Tu: TU \to u(U) \times E$ *is a homeomorphism.*

We leave the proof as an exercise. It is clear that TM with this topology is a Hausdorff space. Moreover, if (U,u) is a chart of M, then (TU,Tu) is a chart of TM over the Banach space $E \times E$, and we have

6.8. PROPOSITION. *The family*

$$\{(TU,Tu);\ (U,u) \text{ is a chart of } M\}$$

defines and analytic structure on TM.

The Banach manifold so constructed on TM is called the *tangent bundle* to M. Obviously, the canonical projections $\pi_1: TM \to M$ and $\pi_2: TM \to E$, given by

$$\pi_1: h \frac{\partial}{\partial u}\Big|_x \to x \quad \text{and} \quad \pi_2: h \frac{\partial}{\partial u}\Big|_x \to h$$

for $h\frac{\partial}{\partial u}\Big|_x \in TM$, are Banach manifold morphisms.

Moreover, if f: M→N is a morphism of Banach manifolds, its derivative df: TM→TN is a morphism of the corresponding tangent bundles.

6.9. DEFINITION. *"An analytic vector field" on a Banach manifold* M *is morphism* X: M→TM *such that we have*

$$\pi_1 \circ X = id_M$$

If X: M→TM is an analytic vector field on M, then its value X(x) at x∈M is a tangent vector to M at x, $X(x) = h(x)\frac{\partial}{\partial u}\Big|_x \in T_xM$. The local expression of X with respect to the charts (U,u) of M and (TU,Tu) of TM is given by

$$u(x) \to (u(x), h) \quad , \quad x \in U$$

where h: M→E is an analytic mapping on M.

We denote by $\mathcal{T}(M)$ the set of all analytic vector fields on M.

6.10. DEFINITION. *Let* $X =: f(x)\frac{\partial}{\partial x}\Big|_x$ *and* $Y = g(x)\frac{\partial}{\partial u}\Big|_x$ *be analytic vector fields on* M *and let* $\lambda \in \mathbb{K}$ *be given. We define*

$$X+Y =: (f(x)+g(x))\frac{\partial}{\partial u}\Big|_x \qquad \lambda X =: \lambda f(x)\frac{\partial}{\partial u}\Big|_x$$

$$[X,Y] =: \left(\frac{\partial g}{\partial u}(x).f(x) - \frac{\partial f}{\partial u}(x).g(x)\right)\frac{\partial}{\partial v}\Big|_x$$

for x∈M.

It is easy to verify that X+Y, λX and [X,Y] are elements of $\mathcal{T}(M)$ and that, in this way, $\mathcal{T}(M)$ becomes a Lie algebra. We call it the Lie algebra of analytic vector fields on M.

6.11. DEFINITION. *Let* ϕ: M→N *be a morphism of Banach manifolds. We say that the analytic vector fields* $X \in \mathcal{T}(M)$ *and* $Y \in \mathcal{T}(N)$ *are "related by* ϕ*" if we have*

$$d\phi . X = Y \circ \phi \tag{6.2}$$

Let us take charts (U,u) of M at x and (V,v) of N at y= ϕ(x), and assume that $X= f(x) \frac{\partial}{\partial u}\Big|_x$ and $Y= g(y) \frac{\partial}{\partial v}\Big|_y$ are the corresponding local expressions of X and Y. Then the expression of (6.2) is given by

$$d\phi(x).f(x) \frac{\partial}{\partial u}\Big|_x = g[\phi(x)] \frac{\partial}{\partial v}\Big|_{\phi(x)} \qquad x\in U$$

6.12. PROPOSITION. *Let* ϕ: M→N *be a morphism of Banach manifolds and assume that* $X_1, X_2 \in \mathcal{T}(M)$ *are related by* ϕ *with* $Y_1, Y_2 \in \mathcal{T}(N)$, *respectively. Then* X_1+X_2, λX_1 *and* $[X_1, X_2]$ *are related by* ϕ *with* Y_1+Y_2, λY_1 *and* $[Y_1, Y_2]$.

We leave the proof as an exercise.

6.13. DEFINITION. *Let* ϕ: M→N *be a morphism of Banach manifolds. Then:*

(a) *We say that* ϕ *is an "immersion" if, for every* x∈M, $d\phi(x): T_xM \to T_{\phi(x)}N$ *is injective and the image* $d\phi(x).T_xM$ *is a closed topologically complemented subspace of* $T_{\phi(x)}N$.

(b) *We say that* ϕ *is a "submersion" if, for every* x∈M, $d\phi(x): T_xM \to T_{\phi(x)}N$ *is surjective and the kernel* Ker $d\phi(x)$ *is a (obviously closed) topologically complemented subspace of* T_xM.

Now we have (see |2| §5).

6.14. PROPOSITION. *Let* ϕ: M→N *be a morphism of Banach manifolds. Then the following statements are equivalent:*

(a) *The mapping* ϕ: M→N *is an immersion and a submersion.*

(b) *For each* x∈M, *the mapping* $d\phi(x): T_xM \to T_{\phi(x)}N$ *is a surjective isomorphism of Banach spaces.*

(c) *For each* x∈M, *there are a neighbourhood* U *of* x *in* M *and a neighbourhood* V *of* y= ϕ(x) *in* N *such that* $\phi|_U$ *is an analytic homeomorphism of* U *onto* V.

6.15. DEFINITION. *If a morphism of Banach manifolds* ϕ: M→N *satisfies any of the above conditions, we say that* ϕ *is a "local isomorphism" of* M *and* N.

By an "isomorphism" of Banach manifolds we mean a bijective local isomorphism ϕ: M→N.

<u>6.16. PROPOSITION</u>. *Let* M, N *and* ϕ *be respectively a topological space, a Banach manifold over* E *and a mapping* ϕ: M→N. *Then the following statements are equivalent*

(a) *For every* x∈M, *there is an open neighbourhood* U *of* x *in* M, *there is a chart* (V,v) *of* y=: ϕ(x) *in* N *and there is a closed topologically complemented subspace* F *of* E *such that* $v\circ\phi$ *is a homeomorphism of* U *onto* $F\cap v[\phi(U)]$.

(b) *There exists a Banach manifold structure on* M *such that its underlying topology is the topology of* M *and* ϕ: M→N *is an immersion.*

The manifold structure satisfying these conditions is unique and its charts are the pairs $(U, v\circ\phi_{|U})$, where U is as in (a). We call it the ϕ-inverse image of the manifold structure in N.

<u>6.17. DEFINITION</u>. *Let* N *be a Banach manifold and denote by* i: M→N *a topological subspace M of N and the canonical inclusion. If the pair* (M,i) *satisfies the conditions of proposition 6.16, we say that* M *endowed with the inverse image manifold structure of that in* N *is a submanifold of* N.

§2.- <u>The concept of a Banach-Lie group</u>.

<u>6.18. DEFINITION</u>. *A "Banach-Lie" group is a set* G *where we have a group structure together with an analytic structure over a Banach space* E *such that the mapping* G×G→G *given by* $(x,y)\mapsto xy^{-1}$ *is analytic.*

According as E is real or complex we say that G is a *real* or a *complex* Banach-Lie group.

If e denotes the identity element of G, we have

<u>6.19. PROPOSITION</u>. *Let the set* G *be endowed with a group structure and an analytic structure over* E. *Then* G *is a Banach-Lie group if and only if the following conditions are satisfied:*

L_1: *For all* $x_0 \in G$, *the mapping* $G \to G$ *given by* $x \to x_0 x$ *is analytic.*

L_2: *For all* $x_0 \in G$, *the mapping* $G \to G$ *given by* $x \to x_0 x x_0^{-1}$ *is analytic in an open neighbourhood of* e.

L_3: *The mapping* $G \times G \to G$ *given by* $(x,y) \to xy^{-1}$ *is analytic and open neighbourhood of* (e,e).

Proof: If G is a Banach-Lie group, then these conditions are obviously satisfied.

Let $(x_0, y_0) \in G \times G$ be given. Then we have

$$xy^{-1} = (x_0 y_0^{-1}) y_0 [(x_0^{-1}x)(y_0^{-1}y)^{-1}] y_0^{-1}$$

for all $x,y \in G$. Thus, the mapping $(x,y) \to xy^{-1}$ can be represented in a neighbourhood of (x_0, y_0) as a composite of mappings of the types mentioned in conditions L_1, L_2 and L_3, whence the result follows.

#

6.20. COROLLARY. *Let* G *be endowed with a group structure and an analytic structure. Then* G *is a Banach-Lie group if and only if the following conditions are satisfied:*

L_1': *The mapping* $G \to G$ *given by* $x \to x^{-1}$ *is analytic on* G.

L_2': The mapping $G \times G \to G$ *given by* $(x,y) \to xy$ *is analytic on* $G \times G$.

Proof: If $(x,y) \to xy^{-1}$ is analytic, so are the mappings $y \to (e,y) \to ey^{-1}$ and $(x,y) \to (x,y^{-1}) \to x(y^{-1})^{-1}$. If $y \to y^{-1}$ and $(x,y) \to xy$ are analytic, so is $(x,y) \to (x,y^{-1}) \to xy^{-1}$.

#

6.21. COROLLARY. *Let* G *be a Banach-Lie group. Then the group structure on* G *is compatible with the topology underlying the analytic structure of* G, *i.e.,* G *is a topological group.*

6.22. EXERCISES. Assume that G is a Banach-Lie group.

Show that the topological group G satisfies the following
conditions:

(a) G is metrizable.

(b) Both the left and right uniform structures of G are
complete.

6.23. DEFINITION. *Let G and H be Banach-Lie groups. We
say that a mapping* f: G→H *is a "morphism" of Banach Lie groups
if f is a morphism of both the group structures and the manifold
structures of G and H.*

6.24. PROPOSITION. *Let G and H be Banach-Lie groups and denote
by* f: G→H *a group homomorphism. Then f is a morphism of Banach-
Lie groups if and only if* f: G→H *is analytic in a neighbourhood
of* e.

Proof: Let $x_0 \in G$ be given. If f: G→H is a group homomorphism, we have $f(x) = f(x_0^{-1}x)$ for all $x \in G$. Thus, by conditions
L_1' and L_2' of corollary 6.20, if f is analytic in a neighbourhood
of e, it is analytic on G. The converse is obvious.

6.25. DEFINITION. *Let the Banach-Lie group G and the
element* $a \in G$ *be given. We define the "left" and "right" transla-
tions" by* a *as the mappings* G→G *given respectively by*

$$\ell_a : x \to ax \qquad\qquad r_a : x \to xa , \qquad x \in G.$$

Obviously, ℓ_a and r_a are automorphisms of the analytic structure of
G. Moreover, the mapping

$$i_a : x \to axa^{-1} \qquad\qquad x \in G$$

is a Banach-Lie group automorphism of G.

6.26. DEFINITION. *Let G be a Banach-Lie group. We say
that a subset* $H \subset G$ *is a "Banach-Lie subgroup" of G is H is a
subgroup and a submanifold of G with respect to the canonical
inclusion* i: H→G.

6.27. EXERCISES. (a) Let G be a Banach-Lie group. Show that the identity component of e is a Banach-Lie subgroup of G.

(b) Let H be a Banach-Lie subgroup of G. Show that H is closed and that the canonical inclusion $i: H \to G$ is a morphism of Banach-Lie groups.

6.28. DEFINITION. *Let* G *be a Banach-Lie group. We say that an analytic vector field* $X \in \mathcal{T}(G)$ *is "left invariant" if, for all* $a \in G$, X *is related to itself by* ℓ_a *, i.e., if we have*

$$(6.3) \qquad d\ell_a \cdot X = X \circ \ell_a \qquad a \in G$$

If (U,u) is a chart of G and $X = f(x) \frac{\partial}{\partial u}\Big|_x$ is the local expression of X, then (6.3) is equivalent to

$$d\ell_a(x) f(x) \frac{\partial}{\partial u}\Big|_x = f[\ell_a(x)] \frac{\partial}{\partial v}\Big|_{ax} \qquad x \in U$$

We denote by $\mathcal{G}(G)$ the subset of $\mathcal{T}(G)$ consisting of all left invariant analytic vector fields on G. As an immediate consequence of proposition 6.12, we obtain

6.29. PROPOSITION. *Let* G *be a Banach-Lie group. Then* $\mathcal{G}(G)$ *is a Lie subalgebra of* $\mathcal{T}(G)$.

6.30. PROPOSITION. *Let* $\alpha: \mathcal{G}(G) \to T_e(G)$ *be the evaluation at the point* $e \in G$. *Then* α *is a surjective isomorphism of vector spaces.*

Proof: Let (U,u) be a chart of G at e; thus α is given by $X \mapsto X(e) \frac{\partial}{\partial u}\Big|_e$ for $X \in \mathcal{G}(G)$. Clearly α is a linear mapping because of definition 6.10. Assume that we have $X(e) = Y(e)$ for some $X, Y \in \mathcal{G}(G)$. As X and Y are left invariant and $a = \ell_a(e)$, we have

$$X(a) = X[\ell_a(e)] = d\ell_a(e).X(e) = d\ell_a(e).Y(e) = Y[\ell_a(e)] = Y(a)$$

for all $a \in G$. Thus $X = Y$ and α is injective.

Let $h \frac{\partial}{\partial u}\Big|_e \in T_e(G)$ be given. Then we define $X(a) =: d\ell_a(e).h \frac{\partial}{\partial u}\Big|_e$

for $a\in G$ and it is immediate to check that $d\ell_e(e): T_e(G)\to T_e(G)$ is the identity mapping. Thus

$$X(e) = d\ell_e(e)\cdot h \frac{\partial}{\partial u}\Big|_e = h \frac{\partial}{\partial u}\Big|_e$$

Moreover, X is analytic. Indeed, since left translations are automorphisms of the manifold structure of G, $(aU, u\circ\ell_a^{-1})$ is a chart of G at the point a and the local expression of ℓ_a is the identity map. Thus, X is locally representable as the constant mapping $x\to h \frac{\partial}{\partial u}\Big|_x$ for $x\in aU$ and X is analytic. Besides, X is left invariant since

$$X[\ell_a(x)] = X(ax) = d\ell_{ax}(e).h \frac{\partial}{\partial u}\Big|_e =$$

$$= d(\ell_a\circ\ell_x)(e)h \frac{\partial}{\partial u}\Big|_e = d\ell_a(x)\left[d\ell_x(e).h \frac{\partial}{\partial u}\Big|_e\right] =$$

$$= d\ell_a(x).\left[d\ell_x(e)\cdot h \frac{\partial}{\partial u}\Big|_e\right] = d\ell_a(x).X(x)$$

for all $a,x\in G$.

Therefore, left invariant vector fields on a Banach Lie group are analytic and they are uniquely determined by their values at the point $e\in G$.

By means of the isomorphism α: $\mathcal{G}(G)\to T_e(G)$ we can transfer the Banach space structure of $T_e(G)$ to $\mathcal{G}(G)$, and it is immediate to verify that, in this way, $\mathcal{G}(G)$ becomes a Banach-Lie algebra. We call it the *Banach-Lie algebra* of G.

§3.- Specific examples: The linear group and its algebraic subgroups.

Let A be a real or complex Banach algebra with unit e. We indistinctly denote by $\mathbb{K}$ any of the fields $\mathbb{R}$ or $\mathbb{C}$.

6.31. DEFINITION. *We define the "commutator product" on* A *by means of*

$$[x,y] =: xy-yx \qquad x,y\in A.$$

It is immediate to see that this product satisfies the conditions of definition 4.20 and that the commutator product $[,]: A\times A\to A$ is continuous. Thus A is a Banach-Lie algebra.

Let us denote by $G(A)$ the set of regular elements of A; thus, $G(A)$ is a group and an open subset of A. Therefore, $G(A)$ is a Banach manifold in a canonical manner (cf. examples 6.3). Now we have

6.32. LEMMA. *With its canonical structures of group and Banach manifold,* $G(A)$ *is a Banach-Lie group whose Banach-Lie algebra is* A.

Proof: It is immediate to check that conditions L_1 and L_2 of proposition 6.19 are satisfied. For $y\in A$ with $\|y-e\|<1$ we have

$$y^{-1} = [e+(y-e)]^{-1} = \sum_{n=0}^{\infty} (-1)^n (y-e)^n$$

the series being convergent in the norm of A. Thus, the mapping $(x,y)\to xy^{-1}$ is analytic in a neighbourhood of (e,e), i.e., condition L_3 is satisfied, too, and $G(A)$ is a Banach Lie group over the Banach space A.

Let $a\in G(A)$ be fixed; with respect to the canonical chart, the expression of the left translation ℓ_a is $\ell_a(x) = ax$ for $x\in G(A)$. Thus, its derivative $d\ell_a$ is given by

$$(6.4) \qquad d\ell_a(x): h \frac{\partial}{\partial u}\Big|_x \to ah \frac{\partial}{\partial u}\Big|_{ax}$$

for all $x\in G(A)$ and $h\in A$. If $X = X(x)\frac{\partial}{\partial u}\Big|$ is a left invariant vector field on $G(A)$, by definition 6.28 we have

$$d\ell_a(x).X(x) = X[\ell_a(x)]$$

By (6.4) this is equivalent to

$$aX(x) = X(ax)$$

for all $a,x\in G(A)$. Taking $x = e$ we obtain $X(a) = aX(e)$ for $a\in G(A)$ or, by changing the notation,

(6.5) $$X(x) \frac{\partial}{\partial u}\Big|_x = xh \frac{\partial}{\partial u}\Big|_x \qquad x\in G(A)$$

where $h=: X(e)\in A$. It is easy to see that vector fields of the form (6.5) are actually left invariant. Moreover, for $X=: xh_1 \frac{\partial}{\partial u}\big|_x$ and $Y=: xh_2 \frac{\partial}{\partial u}\big|_x$ we have

$$[X,Y](x) = (xh_2h_1 - xh_1h_2) \frac{\partial}{\partial u}\Big|_x = x[h_1,h_2] \frac{\partial}{\partial u}\Big|_x$$

so that the mapping $TG(A)\to A$ obtained by evaluating at $e\in G(A)$ is a surjective Banach-Lie isomorphism between $T_eG(A)$ and A.

#

6.33. DEFINITION. *We say that the Banach-Lie group* $G(A)$ *is the "linear group" of the Banach algebra A and denote it by* $GL(A,\mathbb{K})$.

Assume that A is complex Banach algebra; then it is a real Banach algebra, too, which is denote by $A_{\mathbb{R}}$. Thus the linear group $GL(A,\mathbb{C})$, with its underlying real manifold structure, is a real Banach-Lie group over $A_{\mathbb{R}}$. We say that it is the *underlying real linear group* of $GL(A,\mathbb{C})$.

6.34. DEFINITION. *For* $t\in\mathbb{R}$ *and* $x\in A$ *we define*

$$\exp tx =: \sum_{n=0}^{\infty} \frac{t^n}{n!} x^n$$

Since A is complete, we have $\exp tx\in A$; actually, $\exp tx$ is a regular element of A and $(\exp tx)^{-1} = \exp(-tx)$, so that the mapping $\exp: \mathbb{R}\times A\to GL(A,\mathbb{K})$ given by $(t,x)\mapsto \exp tx$ is real analytic. For $t=1$ we simply write $\exp x$ instead of $\exp 1x$. The mapping $A\to GL(A,\mathbb{K})$ given by $x\mapsto \exp x$ is real analytic, too.

6.35. PROPOSITION. *Let A,B be Banach algebras and* $\phi: GL(A,\mathbb{K})\to GL(B,\mathbb{K})$ *a morphism of the corresponding linear groups. Then the derivative* $d\phi(e): T_eGL(A,\mathbb{K})\to T_eGL(B,\mathbb{K})$ *is a homomorphism of the Banach-Lie algebras A and B and we have*

$$\phi(\exp x) = \exp[d\phi(e)x]$$

for all $x \in A$.

Proof: For any fixed $x \in A$, the mapping $f: \mathbb{R} \to GL(B, \mathbb{K})$ given by $t \to f(t) =: \phi(\exp tx)$ is real analytic. Moreover, since ϕ is a group homomorphism, by setting $a =: d\phi(e)x \in B$, we have $f(0) = \phi(e) = e$ and

$$\frac{d}{dt} f(t) = \lim_{s \to 0} \frac{1}{s} [\phi(\exp(t+s)x) - \phi(\exp tx)] =$$

$$= \lim_{s \to 0} \frac{1}{s} \phi(\exp tx)[\phi(\exp sx) - e] =$$

$$= \phi(\exp tx) \lim_{s \to 0} \frac{1}{s} [\phi(\exp sx) - e] = \phi(\exp tx) a$$

Thus, f is the solution of the initial value problem

$$(6.6) \qquad \frac{d}{dt} f(t) = f(t)a \qquad f(0) = e$$

in the Banach space B. Now, we consider the function $g: \mathbb{R} \to GL(B, \mathbb{K})$ given by $t \to g(t) =: \exp ta$ with $a = d\phi(e)x$. It is easy to see that $g(0) = e$ and

$$\frac{d}{dt} g(t) = \sum_{n=1}^{\infty} \frac{t^{n-1}}{(n-1)!} a^n = \left(\sum_{k=0}^{\infty} \frac{t^k}{k!} a^k\right) a = g(t)a$$

Thus g is also the solution of (6.6) and we have $f(t) = g(t)$, i.e. $\phi(\exp tx) = \exp t d\phi(e)x$ for all $t \in \mathbb{R}$ and $x \in A$.

Now, let $x, y \in A$ be given and consider the mapping $F: \mathbb{R} \to GL(B, \mathbb{K})$ given by $F(t) =: \phi(\exp tx)\phi(\exp ty)$, $t \in \mathbb{R}$. A computation similar to the one above gives

$$(6.7) \qquad F'(t) = \phi(\exp tx)(a+b)\phi(\exp ty)$$

where we have put $a = d\phi(e)x$, $b =: d\phi(e)y$. By taking the derivative of (6.7) at $t = 0$

$$F''(0) = a^2 + 2ba + b^2$$

Similarly, if $G(t) =: \phi(\exp ty)\phi(\exp tx)$, we have

$$G''(0) = a^2+2ab+b^2$$

so that $\Psi(t) =: F(t)-G(t)$ satisfies

$$\Psi''(0) = 2[a,b] = 2[d\phi(e)x,\ d\phi(e)y] \tag{6.8}$$

Also, applying twice the chain rule at $t=0$ to compute $\Psi''(0)$ we obtain

$$\Psi''(0) = 2d\phi(e)(yx-xy) = 2d\phi(e)[x,y]$$

whence the result follows by comparing with (6.8).

6.36. EXERCISE. Let $\Psi\colon A\to B$ be a Banach-Lie algebra homomorphism. Show that there exists a unique Banach-Lie homomorphism such that $d\phi(e) = \Psi$.

6.37. PROPOSITION. *Let* H *be a subgroup of the linear group* $GL(A,\mathbb{K})$ *and let* B *be a closed subspace of* A. *Assume that there are a neighbourhood* U *of* e *in* $GL(A,\mathbb{K})$ *and a neighbourhood* V *of* 0 *in* A *such that the exponential mapping* $\exp\colon V\cap B\to U\cap H$ *is a homeomorphism for the corresponding induced topologies. Then* B *is a Banach-Lie subalgebra of* A *and* H *is a Banach-Lie group whose Lie algebra is* B.

Proof: Since $\exp_0^{(1} = \mathrm{id}$, by the inverse mapping theorem, there is no loss of generality in assuming that $(U, \log_{|U})$ is a chart of $GL(A,\mathbb{K})$ at e, where log denotes the inverse of exp. Thus $(U\cap H, \log_{|U\cap H})$ is a chart of H over the Banach space B. Now, let $h\in H$ be given. As H is a subgroup of $GL(A,\mathbb{K})$, the left translation ℓ_h maps $U\cap H$ onto a set $\ell_h(U\cap H)\subset H$ which is a neighbourhood of h in H, and the pair

$$(\ell_h(U\cap H),\ \log \ell_n^{-1}|_{\ell_h(U\cap H)}) \tag{6.9}$$

is a chart of H at h. It is easy to see that the family given by (6.9) for $h\in H$ is an analytic structure over B. Moreover, for this analytic structure, H is a Banach-Lie group; thus by proposition 6.30, B is a Lie subalgebra of A. #

6.38. REMARK. Notice that, in general, H is not a Banach-Lie subgroup of $GL(A,\mathbb{K})$ because, as a manifold, H may fail to be a submanifold of $GL(A,\mathbb{K})$: its tangent space at e is B, which in general is not a complemented subspace of A.

6.39. DEFINITION. *Let A be a Banach algebra over* $\mathbb{K}$ *and let* H *be a subgroup of* $GL(A,\mathbb{K})$. *We say that* H *is a* $\mathbb{K}$*-algebraic subgroup of degree* $\leq n$ *of* $GL(A,\mathbb{K})$ *if there exists a non void set* S *of continuous vector-valued* $\mathbb{K}$*-polynomials* $q: A\times A\to E$ *of degree* $\leq n$ *with* $q(0,0)=0$ *such that we have*

$$(6.10)\qquad H=\{z\in GL(A,\mathbb{K});\quad q(z,z^{-1})=0\qquad \forall q\in S\}$$

Of particular interest for us, though not included in the above definition, is the situation in which we have a Banach algebra over $\mathbb{C}$ and a subgroup H of $GL(A,\mathbb{C})$, but the equations (6.10) defining H are $\mathbb{R}$-polynomials $q: A\times A\to E$ on the underlying $\mathbb{R}$-structures of $A\times A$ and E

Notice that, in all these cases, H is closed in $GL(A,\mathbb{K})$. Clearly, any finite product and any intersection of $\mathbb{K}$-algebraic subgroups of degree $\leq n$ is a $\mathbb{K}$-algebraic subgroup of degree $\leq n$. The definition includes the case in which H is defined by a set S of $\mathbb{K}$-polynomials $q: A\to E$ depending on a single variable. Also, by the Hanh-Banach theorem, the polynomials $q\in S$ can be chosen to be $\mathbb{K}$-valued.

6.40. THEOREM. *Let A be a Banach algebra over* $\mathbb{K}$ *and assume that* H *is a* $\mathbb{K}$*-algebraic subgroup of degree* $\leq n$ *of* $GL(A,\mathbb{K})$. *Then* H *is a Banach Lie group whose Banach-Lie algebra is*

$$B=:\{y\in A;\ \exp ty\in H\qquad \forall t\in\mathbb{R}\}$$

Proof: We write $w=:(u,v)$ for the elements of $A\times A$ which is a Banach algebra over $\mathbb{K}$ with respect to the norm $\|w\|=:\max\{\|u\|,\|v\|\}$. Let us put $P=:\bigoplus_{k=1}^{n} P_k(A\times A)$ for the $\mathbb{K}$-Banach space of continuous $\mathbb{K}$-polynomials $p: A\times A\to\mathbb{K}$ of degree $\leq n$ such that $p(0,0)=0$. Now we define a mapping

Φ: GL(A,$\mathbb{K}$)$\to$GL(P(P),$\mathbb{K}$) by means of

$$[\Phi(x)p](u,v)=: p(ux,x^{-1}v)$$

where p$\in$P, (u,v)$\in A\times A$ and x$\in$GL(A,$\mathbb{K}$). Also, we define a mapping Ψ: $A\to L$(P) by

$$[\Psi(x)p](u,v)=: p^{(1}_{(u,v)}(ux,-xv)$$

where p$\in$P, (u,v)$\in A\times A$ and x$\in A$.

First, we study some properties of Φ and Ψ. We have

(a) ϕ is a Banach-Lie group homomorphism and dϕ(e)= Ψ. The proof is an exercise. Thus, by proposition 6.35 we get

(b) Ψ is a Banach-Lie algebra homomorphism and

$$[\phi(\exp x)p]= [\exp\psi(x)]p \tag{6.11}$$

for all x$\in$A and p$\in$P.

(c) Let z$\in$A be given. Then each of the subspaces P_k(A×A), k= 1,2,..,n, is invariant by Ψ(z), i.e., we have $\Psi(z)\subset P_k$(A×A). Moreover, if z$\in$A is a regular element of A, then Ψ(z) is a regular element of L(P).

Indeed, let p$\in P$(A×A) be given and suppose that F$\in L_k(A\times A,\mathbb{K})$ is its associated symmetric k-linear mapping so that we have

$$p(w)= F(w,w,..,w) \qquad w\in A\times A$$

Then, the mapping f: A×A$\to$A×A given by

$$f(z): w= (u,v) \to (uz,-zv) \tag{6.12}$$

satisfies f(z)$\in L$(A×A). Therefore, from the definition of ψ we get

$$[\Psi(z)p]w= kF(f(z)w; w,..,w) \qquad w\in A\times A$$

and Ψ(z)p$\in P_k$(A×A).

Now, suppose that z$\in$A is regular. Then f(z) as defined by (6.12) is regular in L(A×A) and $f(z)^{-1}= f(z^{-1})$. Thus, the restriction of $\Psi(z)$ to each of the subspaces $P_k(A\times A)$ is a regular element of $L(P_k(A\times A))$, the inverse image of $p\in P_k(A\times A)$ being given by

$$[\Psi(z)^{-1}p]w = \frac{1}{k} F(f(z)^{-1}w; w,..,w) \qquad w\in A\times A$$

Therefore $\Psi(z)$ is regular in $L(P)$, too.

Next we show that B is a closed Lie subalgebra of A.

Since H is a $\mathbb{K}$-algebraic subgroup of degree $\leq n$ of $GL(A,\mathbb{K})$, there is a set of $\mathbb{K}$-polynomials $S\subset P$ such that

$$H = \{z\in GL(A,\mathbb{K});\ \ q(z,z^{-1}) = 0 \qquad \forall q\in S\}$$

We define another set of polynomial $Q\subset P$ by means of

$$(6.13) \qquad Q =: \{p\in P;\ \ p(h,h^{-1}) = 0 \qquad \forall h\in H\}$$

Clearly, Q is a closed $\mathbb{K}$-subspace of P and $S\subset Q$; thus, in particular

$$(6.14) \qquad [z\in GL(A,\mathbb{K}),\ q(z,z^{-1}) = 0 \quad \forall q\in Q] \Rightarrow \quad z\in H$$

We claim that, for $x\in GL(A,\mathbb{K})$, we have the equivalence

$$(6.15) \qquad x\in H \iff \phi(x)Q\subset Q$$

Indeed, let $x\in GL(A,\mathbb{K})$ be given and assume that $x\in H$. As H is a subgroup of $GL(A,\mathbb{K})$, we have $Hx = H^{-1}x = H$. From (6.13) and the definition of ϕ we obtain

$$[\phi(x)q](h,h^{-1}) = q(hx,x^{-1}h^{-1}) = 0$$

for all $q\in Q$ and $h\in H$; thus $\phi(x)Q\subset Q$ by (6.13). Conversely, let $x\in GL(A,\mathbb{K})$ be given and assume that $\phi(x)Q\subset Q$. By (6.14) it suffices to show that $q(x,x^{-1}) = 0$ for all $q\in Q$. Let $q\in Q$ be given; by assumption we have $\phi(x)q\in Q$; thus by (6.13) we obtain $[\phi(x)q](h,h^{-1}) = 0$, i.e., $q(hx,x^{-1}h^{-1}) = 0$ for all $h\in H$, and taking $h = e\in H$ we get $q(x,x^{-1}) = 0$.

Now we claim that, for $y\in A$, we have the equivalence

(6.16) $$y \in B \iff \Psi(y)Q \subset Q$$

Indeed, let $y \in B$ be given. Then we have $\exp ty \in H$ for all $t \in \mathbb{R}$ and, by (6.15), $\phi(\exp ty)Q \subset Q$; therefore, from (6.11) we derive $[\exp \Psi(y)]Q \subset Q$. If we fix any $q \in Q$, the mapping $\mathbb{R} \to P$ given by $t \to \exp \Psi(y) q$ takes its values in the closed $\mathbb{K}$-subspace Q of P and, by taking its derivative at $t = 0$, we get $\Psi(y)q \in Q$, whence $\Psi(y)Q \subset Q$. Conversely, let $y \in A$ be such that $\Psi(y)Q \subset Q$. As Q is a closed $\mathbb{K}$-subspace of P, we have $[\exp t\psi(y)]Q \subset Q$ for all $t \in \mathbb{R}$. Thus, by (6.11), $\phi(\exp ty)Q \subset Q$, whence $\exp ty \in H$ for all $t \in \mathbb{R}$ and therefore $y \in B$.

In particular, as $d\phi(e) = \Psi$ is a Banach-Lie algebra homomorphism, (6.16) entails that B is a closed Lie subalgebra of A.

Next, we show that H is a Banach-Lie group.

Let $A^{\mathbb{C}} =: A \oplus iA$ be the complexified of the Banach algebra A (thus $A^{\mathbb{C}} = A$ when $\mathbb{K} = \mathbb{C}$). For $x \in A^{\mathbb{C}}$, $Sp(x)$ is the spectrum of x in $A^{\mathbb{C}}$. From the spectral theory we know that the sets

$$U =: \{x \in GL(A,\mathbb{K});\ Sp(x) \subset \{\lambda \in \mathbb{C};\ |\arg\lambda| < \pi/n\}\}$$

and

$$V =: \{y \in A;\ Sp(y) \subset \{\mu \in \mathbb{C};\ |\mathrm{img}\,\mu| < \pi/n\}\}$$

are, respectively, neighbourhoods of e in $GL(A,\mathbb{K})$ and 0 in A. According to the holomorphic functional calculus (c.f. |1|), on U we can select a holomorphic branch of the logaritmic function. Let us denote by *log* its principal determination. By the spectral mapping theorem, $\log: U \to V$ is a complex bianalytic map (thus, a real bianalytic map, too, in case $\mathbb{K} = \mathbb{R}$) whose inverse is $\exp: V \to U$. Therefore, by proposition 6.37, it suffices to show that we have

$$\exp(V \cap B) \subset U \cap H \qquad \log(U \cap H) \subset V \cap B$$

Now, let $y \in V \cap B$ be given; from $y \in V$ and $y \in B$ we get $\exp y \in U$ and $\exp ty \in H$ for all $t \in \mathbb{R}$, thus $\exp y \in U \cap H$.

Next, let $x\in U\cap H$ be given and put $y=: \log x$. Thus, in particular

$$y\in V \tag{6.17}$$

Moreover, from $x\in U$ we derive

$$Sp(x)\subset\{\lambda\in\mathbb{C};\ |\arg\lambda|<\pi/n\}$$

and by the spectral mapping theorem

$$Sp(y)= Sp(\log x)\subset\{\mu\in\mathbb{C};\ \ |\mathrm{img}\,\mu|<\pi/n\} \tag{6.18}$$

We claim that the spectrum $Sp[\Psi(y)]$ of $\Psi(y)$ in the complexified algebra $\mathcal{L}(P)^{\mathbb{C}}$ of $\mathcal{L}(P)$ (or in $\mathcal{L}(P)$ when $\mathbb{K}= \mathbb{C}$) satisfies

$$Sp[\Psi(y)]\subset\{\lambda\in\mathbb{C};\ \ |\mathrm{img}\,\lambda|<\pi\}$$

Indeed, we prove that for $\lambda\in\mathbb{C}$ with $|\mathrm{img}\,\lambda|\geqslant\pi$, $\lambda I-\Psi(y)$ is a regular element of $\mathcal{L}(P)^{\mathbb{C}}$. Now we have

$$(\lambda I-\Psi(y))p= \lambda p-\Psi(y)p$$

so that

$$[\lambda p-\Psi(y)p]w= \lambda F(w,w,..,w)-kF(f(y)w;w,..,w) = kF[(\frac{\lambda}{k}I-f(y))w;w,..,w]$$

for $p\in P_k(A\times A)$, $k= 1,2,..,n$, and $w= (u,v)\in A\times A$. Thus, it suffices to prove that $\frac{\lambda}{k}I-f(y)$ is regular in $\mathcal{L}(A\times A)$ for $k= 1,2,..,n$. But, due to the definition of $f(y)$,

$$[\frac{\lambda}{k}I-f(y)](u,v)= [u.(\frac{\lambda}{k}e-y),\ (\frac{\lambda}{k}e+y)v]$$

Since by (6.18), $|\mathrm{img}\,\lambda|\geqslant\pi/n$ entails $\lambda/k\notin Sp(y)$ and $\lambda/k\notin Sp(-y)$ for $k= 1,2,..,n$, the elements $\frac{\lambda}{k}e-y$ and $\frac{\lambda}{k}e+y$ are regular in A and so is $\frac{\lambda}{k}I-f(y)$ in $\mathcal{L}(A\times A)$.

Since $\phi(x)= \phi(\exp y)= \exp\Psi(y)$, by the functional calculus and the spectral mapping theorem it follows that

$$\Psi(y)= \log\phi(x)\quad \text{and}\quad Sp[\phi(x)]\subset\{\lambda\in\mathbb{C};\ |\arg\lambda|<\pi\}$$

By Runge's theorem, there is a sequence of polynomials $p_k \in \mathbb{C}|\lambda|$, $k \in \mathbb{N}$, such that we have $\log\lambda = \lim_{k\to\infty} p_k(\lambda)$ uniformly when $\lambda \in Sp|\phi(x)|$. Then

$$\Psi(y) = \log\phi(x) = \frac{1}{2\pi i}\int_\gamma [\lambda e - \phi(x)]^{-1}\log\lambda d\lambda =$$

$$= \frac{1}{2\pi i}\int_\gamma \lim_{k\to\infty} p_k(\lambda)[\lambda e - \phi(x)]^{-1}d\lambda = \lim_{k\to\infty}\frac{1}{2\pi i}\int_\gamma p_k(\lambda)[\lambda e - \phi(x)]^{-1}d\lambda =$$

$$= \lim_{k\to\infty} p_k[\phi(x)]$$

Since $x \in H$, by (6.15) we have $\phi(x)Q \subset Q$ and, as Q is a closed subspace of P, we get $\Psi(y)Q = \lim_{k\to\infty} p_k[\phi(x)]Q \subset Q$. Then, by (6.16) we obtain

$$\Psi(y) \in B \tag{6.19}$$

Finally, from (6.17) and (6.19) we deduce $\Psi(y) \in V \cap B$.

#

<u>6.41. REMARK</u>. The case in which A is a Banach algebra over $\mathbb{C}$ and H is an $\mathbb{R}$-algebraic subgroup of $GL(A,\mathbb{C})$ can also be included in our considerations. Indeed, we can consider the underlying $\mathbb{R}$-structures of A and $GL(A,\mathbb{C})$ and define $P = \bigoplus_{k=1}^{n} P_k(A\times A)$ to be the Banach space of continuous $\mathbb{R}$-polynomials $p: A\times A \to \mathbb{C}$ of degree $\leqslant n$ with $p(0,0) = 0$. As in our case the polynomials defining H belong to a subset of P, we are in a situation in which theorem 6.40 is applicable.

A number of interesting examples of $\mathbb{K}$-algebraic subgroups of degree $\leqslant n$ of $GL(A,\mathbb{K})$ are included in the following

<u>6.42. PROPOSITION.</u> *Let* X, Y *and* $f \in L(^nX,Y)$ *be respectively two Banach spaces over* $\mathbb{K}$ *and a* $\mathbb{K}$*-multilinear mapping* $f: X\times\ldots\times X \to Y$. *Let* $m = 0$ *or* $m = 1$ *and suppose that* $X = Y$ *when* $m = 1$. *Then, the set* H *of the elements* $\alpha \in GL(L(X),\mathbb{K})$ *satisfying*

(6.20) $f(\alpha x_1,..,\alpha x_n) = \alpha^m f(x_1,..,x_n) \qquad x_1,..,x_n \in X$

is a Banach-Lie group whose Banach-Lie algebra B *is the set of all* $\delta \in L(X)$ *satisfying*

(6.21) $f(\delta x_1,..,x_n)+..+f(x_1,...,\delta x_n) = m\delta f(x_1,..,x_n)$

Here α^m and $m\delta$ denote respectively the identity and the zero transformation on Y when m= 0.

Proof: Obviously $A =: L(X)$ is a Banach algebra over $\mathbb{K}$ and H, as defined by (6.20) is a subgroup of the linear group $GL(L(X),\mathbb{K})$. We claim that H is a $\mathbb{K}$-algebraic subgroup of degree $\leq n$. Indeed, for fixed $x_1,x_2,..x_n \in X$, the mapping $p_{x_1,..,x_n} : L(X) \to X$ given by

$$p_{x_1,..,x_n}(\alpha) =: f(\alpha x_1,..,\alpha x_n) - \alpha^m f(x_1,..,x_n), \qquad \alpha \in L(X)$$

is a continuous n-homogeneous $\mathbb{K}$-polynomial, and H is defined by the set S of equations

$$p_{x_1,..,x_n}(\alpha) = 0$$

for $x_1,..,x_n \in X$. Moreover, B as defined by (6.21), is a closed Banach-Lie subalgebra of $A = L(X)$. Thus, it suffices to show that H and B are as in theorem 6.40 i.e. we have

$$B = \{\alpha \in L(X);\ \exp t\alpha \in H \quad \forall t \in \mathbb{R}\}$$

Now, suppose that $\delta \in L(X)$ satisfies $\exp t\delta \in H$ for all $t \in \mathbb{R}$. Replacing α by $\exp t\delta$ in (6.20) and taking the derivative at t= 0 we see that δ satisfies (6.21); thus $\delta \in B$. Conversely, suppose that $\delta \in B$ and define linear mappings $F_0,F_1,..,F_n : L(^nX,Y) \to L(X,Y)$ by means of

$$F_0(G)(x\ ,..,x\ ) = mG(x_1,..,x_n)$$

$$F_k(G)(x_1,..,x_n) = G(x_1,..,x_{k-1},\delta x_k,x_{k+1},..,x_n)$$

for $G\in L(^nX,Y)$, $x_1,..,x_n\in X$ and $k= 1,2,..,n$. Let us write

$$F=: (F_1+F_2+..+F_n)-F_0$$

Since δ satisfies (6.21) we have $F(f)= 0$ and $F_0,F_1,..,F_n$ commute. Therefore

$$(\exp F_1)..(\exp F_n)f= \exp(F_1+..+F_n)f= (\exp F_0)(\exp F_n)f=(\exp F_0)f=$$
$$= (\exp\delta)^m f$$

which shows that $\exp\delta\in H$. Since B is a linear space, we can do the same argument with $t\delta$ instead of δ, whence we conclude that $\exp t\delta\in H$ for all $t\in\mathbb{R}$.

#

6.43. EXAMPLES. Let A be a Banach algebra over $\mathbb{K}$ (where A may have no unit and fail to be associative, for example, any Banach-Lie algebra of any Banach-Jordan algebra). Then we can apply proposition 6-42 to the case in which $n= 2$, $m= 1$, $X= Y= A$ and $f\in L(^2X,X)$ is the multiplication on A, i.e., $f(x,y)= x.y$. Observe that the auxiliary Banach algebra $L(X)$ appearing in proposition 6.42 is now $L(A)$ which is associative and has unit even if A fails to be so. Thus, theorem 6.40 and proposition 6.42 are applicable. Accordingly, the set

$$H= \{\alpha\in GL(L(A),\mathbb{K});\ \alpha(xy)= \alpha(x)\alpha(y) \quad \forall x,y\in A\}$$

which is the group of automorphisms of A, is a Banach-Lie group in the normed topology of $L(A)$. The Banach-Lie algebra of this group is

$$B= \{\delta\in L(A);\ (\delta x).y+x.(\delta y)=\delta(x.y) \ \ \forall x,y\in A\}$$

which is the algebra of $\mathbb{K}$-derivations of A. Besides, H is the $\mathbb{K}$-algebraic subgroup of $GL(L(A),\mathbb{K})$ of degree ≤ 2 defined by the equations $p_{x,y}(\alpha)= 0$ where

$$p_{x,y}(\alpha)= \alpha(x).\alpha(y)-\alpha^2(x.y)$$

and $x,y\in A$.

Let X be a complex Hilbert space and $Y=\mathbb{C}$. Then can apply proposition 6.42 with $n=2$, $m=0$ to the spaces X and Y and the real bilinear mapping $f\in L(^2X,\mathbb{C})$ given by the scalar product on X, $f(x,y)=(x|y)$. Accordingly, if α^* denotes the adjoint of the operator $\alpha\in L_{\mathbb{R}}(X)$, the set

$$H=:\{\alpha\in GL(L_{\mathbb{R}}(X),\mathbb{C});\ (\alpha x|\alpha y)=(x|y),\ \alpha(ix)=i\alpha(x)\quad \forall x,y\in H\}=$$

$$=\{\alpha\in GL(L_{\mathbb{C}}(X),\mathbb{C});\ \alpha^*\alpha=I\},$$

which is the *unitary group* of H, is a real Banach-Lie group whose Banach-Lie algebra is

$$B=\{\delta\in L_{\mathbb{C}}(X);\ (\delta x|y)+(x|\delta y)=0\quad \forall x,y\in X\}=\{\delta\in L_{\mathbb{C}}(X);\ \delta+\delta^*=0\}$$

The unitary group of X is a real algebraic subgroup of $GL(L_{\mathbb{C}}(X),\mathbb{K})$ of degree ≤ 2 defined by the equations $p_{x,y}(\alpha)=0$ where

$$p_{x,y}(\alpha)=(\alpha x|\alpha y)-(x|y)$$

and $x,y\in X$.

Let X be a Hilbert space over $\mathbb{C}$ and denote by Q any conjugation on X. Take $f\in L(^2X,\mathbb{C})$ to be the real bilinear mapping given by $f(x,y)=:(Qx|y)$, $x,y\in X$. Then we can apply proposition 6.42 with $n=2$, $m=0$. If α^t denotes the transposed of the operator $\alpha\in L_{\mathbb{R}}(X)$, then, the set

$$H=:\{\alpha\in Gl(L_{\mathbb{R}}(C),\mathbb{C});\ (Q\alpha x|\alpha y)=(Qx|y),\quad \alpha(ix)=i\alpha(x)\quad \forall x,y\in X\}=$$

$$=\{\alpha\in GL(L_{\mathbb{C}}(X),\mathbb{C});\ \alpha^t\alpha=I\},$$

which is the *orthogonal group* of X, is a real Banach-Lie group whose Banach-Lie algebra is

$$B=\{\delta\in L_{\mathbb{C}}(X);\quad (Q\delta x|y)+(Qx|\delta y)=0\quad \forall x,y\in X\}=\{\delta\in L_{\mathbb{C}}(X);\ \delta+\delta^t=0\}$$

The orthogonal group of X is a real algebraic subgroup of

$GL(L_{\mathbb{C}}(X),\mathbb{C})$ of degree ≤ 2 defined by the equations $p_{x,y}(\alpha)=0$ where

$$p_{x,y}(\alpha)=(Q\alpha x|y)-(Qx|y)$$

and $x,y\in X$.

<u>6.44. EXERCISES</u>. Let X,Y be Hilbert spaces over $\mathbb{C}$. A J^*-algebra is a closed complex subspace $\mathcal{U}$ of $L_{\mathbb{C}}(X,Y)$ such that we have $AB^*A\in\mathcal{U}$ whenever $A,B\in\mathcal{U}$. A J^*-automorphism is any $\alpha\in GL(L_{\mathbb{C}}(\mathcal{U}),\mathbb{C})$ such that we have

$$\alpha(AB^*A)=\alpha(A)\alpha(B)^*\alpha(A)$$

for all $A,B\in\mathcal{U}$. Prove that the group of all J^*-automorphisms of $\mathcal{U}$ is a real algebraic subgroup of $GL(L_{\mathbb{C}}(\mathcal{U}),\mathbb{C})$ of degree ≤ 3. Thus it is a real Banach-Lie group. Prove that its Banach-Lie algebra is the set

$$B=\{\delta\in L_{\mathbb{C}}(\mathcal{U});\ (\delta A)B^*A+A(\delta B)^*A+AB^*(\delta A)=0\quad \forall A,B\in\mathcal{U}\}$$

of all J^*-derivations of $L_{\mathbb{C}}(\mathcal{U})$.

§4.- <u>Local behaviour of the exponential map at the origin</u>.

Now we turn our attention to the construction of a real Banach-Lie group structure on AutD.

Throughout this section, $B\subset D$ and $\delta>0$ stand for a fixed open ball and a real number such that $B_\delta\subset\subset D$. We know that $(\text{autD},\|\cdot\|_{B_\delta})$ is a real Banach-Lie algebra. We denote by $Hol_\infty(B,E)$ the complex Banach space of holomorphic mappings $f\colon B\to E$ that are bounded on B. Thus $(Hol_\infty(B,E),\|\cdot\|_B)$ is a real Banach space, too.

As autD and $Hol_\infty(B,E)$ will always be endowed with the topologies respectively defined on them by the norms $\|\cdot\|_{B_\delta}$ and $\|\cdot\|_{B}$, we shall omit any reference to these norms. However, we shall consider several topologies on AutD $Hol_\infty(B,E)$; thus, in order to avoid any possible confusion, whenever we refer to AutD we shall explicitely mention the topology we are considering on it.

6.45. LEMMA. *There is a neighbourhood* M *of the origin in* autD *such that, for any* $A\in M$, *the series*

$$\sum_{n=0}^{\infty} \frac{1}{n!} \hat{A}^n id_D$$

is convergent to $(\exp A)_{|B}$ *in the space* $Hol_\infty(B,E)$. *The mapping* $\exp: M \to Hol_B(B,E)$ *given by*

$$A \to (\exp A)_{|B} = \sum_{n=0}^{\infty} \frac{1}{n!} \hat{A}^n id_D \tag{6.22}$$

is real analytic on M .

Proof: Let $A\in autD$ be given. Then, for $t\in\mathbb{R}$ and $n\in\mathbb{N}$, we have $\frac{t^n}{n!} \hat{A}^n id_D \in Hol_\infty(B,E)$; therefore, we can define a formal power series $\mathbb{R}\to Hol_\infty(B,E)$ by means of

$$t \to \sum_{n=0}^{\infty} \frac{t^n}{n!} \hat{A}^n id_D \tag{6.23}$$

As in the proof of proposition 4.1, we have

$$\left\| \frac{1}{n!} \hat{A}^n id_D \right\|_B \leq M\left(\frac{2e}{\delta} \| A \|_{B_\delta} \right)^n$$

for all $n\in\mathbb{N}$, where $M =: \| id_D \|_B < \infty$ is independent of n. Now,

$$M =: \{A\in autD;\ \| A \|_{B_\delta} < \frac{\delta}{2e}\}$$

is a neighbourhood of the origin in autD and, for any fixed $A\in M$, the radius of convergence of (6.23) is greater than 1. Since $Hol_\infty(B,E)$ is complete,

$$f(t,A)_{|B} =: \sum_{n=0}^{\infty} \frac{t^n}{n!} \hat{A}^n id_B$$

is convergent in $Hol_\infty(B,E)$ to $(\exp tA)_{|B}$ for all $t\in[-1,+1]$. Moreover, it is easy to see that

$$\frac{d}{dt} f(t,A) = \sum_{n=1}^{\infty} \frac{t^{n-1}}{(n-1)!} \hat{A}^n id_D = A[f(t,a)]$$

and $f(0,A) = id_D$. Thus, by definition 4.4,

$$f(1,A) = \sum_{n=0}^{\infty} \frac{1}{n!} \hat{A}^n id_B = (\exp A)|_B$$

Now (6.22) defines a formal power series between the real Banach spaces autD and $Hol_\infty(B,E)$. Since for $A \in M$ this series is convergent, exp: $M \to Hol_\infty(B,E)$ defines a real anlaytic mapping on M.

#

<u>6.46. REMARK.</u> Notice that exp: $M \to Hol_\infty(B,E)$ takes its values not only in the space $Hol_\infty(B,E)$ but in the smaller set AutD.

<u>6.47. LEMMA</u>. *There are a neighbourhood* M *of* 0 *in* autD *and a neighbourhood* N *of* id_D *in* $Hol_\infty(B,E)$ *such that* exp $M \to N$ *is a bijection. Moreover, both* exp: $M \to N$ *and its inverse* log: $N \to M$ *are lipschitzian on* M *and* N .

Proof: By lemma 6.45 exp: $M \to Hol_\infty(B,E)$ is a real analytic mapping on a neighbourhood M of 0 in autD. Its derivative at the origin is the element of $L(\text{autD}, Hol_\infty(B,E))$ given by

$$\exp_0^{(1}A = \hat{A}\, id_D = A$$

Thus, by the inverse mapping theorem, there are a nieghbourhood M' of 0 in autD and a neighbourhood N' of id in $Hol_\infty(B,E)$ such that exp: $M' \to N'$ is a bianalytic mapping. By the continuity of the derivative at 0, there is a convex neighbourhood M'' of 0 in autD in which $\exp^{(1}$ is bounded

$$K =: \sup\{ \| \exp_A^{(1} \| \; ; \; A \in M''\} < \infty$$

Then, for $A_1, A_2 \in M''$ we have

$$\| \exp A_1 - \exp A_2 \|_B \leq \| A_1 - A_2 \|_{B_\delta} \sup\{ \| \exp^{(1}_{A_1 + tA_2} \| \; ; \; 0 \leq t \leq 1\} \leq$$

$$\leq K \| A_1 - A_2 \|_{B_\delta}$$

and exp is lipschitzian on M''. A similar argument applies to its inverse log. There is no loss of generality in assuming that $M''= M$ and $N= \exp M$.

6.48. QUESTION. Is $N=: \exp M$ a T-neighbourhood of id_D in AutD?.

6.49. EXERCISE. Show that if question 6.48 has an affirmative answer, then by lemma 6.47, every $F\in$AutD admits a neighbourhood that is homeomorphic to M by $A\to F \exp A$.

However, as we shall see in chapter 8, the answer is not always affirmative. Thus, in general, we may only expect that for some group topology, which is finer than T, the mappings $A\to\exp A$, $A\in M$, are local homeomorphisms of AutD over autD for all $F\in$AutD.

To establish the existence of such topology we should know that the composite mapping $\exp A_1 \circ \exp A_2$ can always be written in the form exp C for some $C\in$autD, whenever A_1, A_2 are sufficiently near to 0 in autD. This fact is a special case of one of the the main goals of the general Lie theory, known as the Campbell-Hausdorff theorem (cf. |3|).

6.50. THEOREM. *There are a neighbourhood M of the origin in* autD *and a real analytic mapping* $C: M\to$autD *such that we have*

$$\exp C(A_1,A_2) = \exp A_1 . \exp A_2 \tag{6.24}$$

for all $A_1,A_2\in M$.

By the continuity of C at the origin, we can find neighbourhoods $M_1\subset M$ and $M_2\subset M$ of 0 in autD such that

$$C(M_1, M_2)\subset M \quad \text{and} \quad C(M_2,M_2) \subset M_1 \tag{6.25}$$

6.51. REMARK. The explicit form of the mapping C is also known. One can show (cf. |3|) that given $B\subset\subset D$, we have

$$C(s_1A_1,\ s_2A_2)|_B = C(s_1A_1,\ s_2A_2)\hat{}\ \mathrm{id}_B =$$
$$= \log(\exp s_1\hat{A}_1,\ \exp s_2\hat{A}_2)\,\mathrm{id}_B$$

in the sense that the formal power series

$$\sum_{k,\ell\geq 0} s_1^k s_2^\ell X_{k,\ell}(A_1,A_2) \sim \log\left[id+(\exp s_1\hat{A}_1 \exp s_2\hat{A}_2-id)\right] ,$$

where

$$X_{k,\ell}(A_1,A_2) =:$$

$$=: \sum_{n=1}^{k+1} (-1)^{n+1} \sum_{\substack{p_1+..+p_n=k,\ q_1+..+q_n=\ell \\ p_i q_i\geq 0,\quad p_i+q_i>0}} \frac{1}{p_1!q_1!} \cdots \frac{1}{p_n!q_n!} \hat{A}_1^{p_1}\hat{A}_2^{p_2}\cdots$$

$$\cdots\hat{A}_1^{p_n}\hat{A}_2^{q_n}$$

converges in the norm $\|\cdot\|_B$ to id_B whenever $\|s_1A_1\|$ and $\|s_2A_2\|$ are sufficiently small. (This is not consequence of any majorization!). Then, we necessarily have

$$C(A_1,A_2) = \sum_{k,\ell} C_{k,\ell}(A_1,A_2)$$

where

$$C_{k\ell}(A_1,A_2) =: X_{k,\ell}(A_1,A_2)\,id_D$$

and the convergence is meant in the topology of autD. Since $C_{k,\ell}(A_1,A_2)\in$autD for all k,ℓ because they are partial derivatives in the T-sense of the mapping $(s_1,s_2)\to C(s_1A_1,s_2A_2)$ at 0, Dynkin's identity yields

$$C_{k,\ell}(A_1,A_2) =$$

$$= \sum_{n=1}^{k+1} (-1)^{n+1} \frac{1}{p_1!q_1!} \cdots \frac{1}{p_n!q_n!} A_{1\#}^{p_1}A_{2\#}^{q_1}\cdots A_{1\#}^{p_{n-1}}A_{2\#}^{q_{n-1}}A_{1\#}^{p_n}A_2^{q_{n-1}}A$$

It would be interesting to have a direct proof for the formula

$$\exp A_1 . \exp A_2 = \exp\left[\sum_{k,\ell} C_{k,\ell}(A_1,A_2)\right]$$

in the setting of AutD.

6.52. THEOREM. *There exists a unique Hausdorff topology T_a on* AutD *such that* (AutD, T_a) *is a topological group and*

$$\{\exp \frac{1}{n} M;\ n = 1,2,..\}$$

is a fundamental system of neighbourhooods of id_D *for* T_a. *Moreover,* $T_a \geqslant T$.

Proof: From the general theory of topological groups, it suffices to prove that the statements (a), (b), (c) and (d) below are satisfied.

(a) We have $\bigcap_{n=1}^{\infty} \exp \frac{1}{n} M = \{id_D\}$.

Indeed, let $f \in \exp M$ be such that $f \neq id_D$. Then, there is some $A \in M$ with $A \neq 0$ for which $\exp A = f$; therefore, we can find some $n \in \mathbb{N}$ such that $A \notin \frac{1}{n} M$. Since the exponential mapping is injective on M, we have $f \notin \exp \frac{1}{n} M$; thus $f \notin \bigcap_{n=1}^{\infty} \exp \frac{1}{n} M$.

(b) Let n_1 and $n_2 \in \mathbb{N}$ be given; then there exists some $m \in \mathbb{N}$ such that $\exp \frac{1}{m} M \subset (\exp \frac{1}{n_1} M) \cap (\exp \frac{1}{n_2} M)$.

Indeed, it suffices to consider $m =: \max(n_1, n_2)$.

(c) Let $n \in \mathbb{N}$ be given; then there exists some $m \in \mathbb{N}$ such that $(\exp \frac{1}{n} M).(\exp \frac{1}{n} M)^{-1} \subset \exp \frac{1}{m} M$.

Indeed, by (6.24) we have $C(0,0) = 0$. As C is continuous at 0, given $n \in \mathbb{N}$ we can find $m \in \mathbb{N}$ such that $C(\frac{1}{m} M, \frac{1}{m} M) \subset \frac{1}{n} M$. We may assume M to be symmetric, i.e., $M = -M$. Then

$$(\exp \frac{1}{m} M)(\exp \frac{1}{m} M)^{-1} = (\exp \frac{1}{m} M).(\exp \frac{-1}{m} M) =$$

$$= \exp[C(\frac{1}{m} M, \frac{1}{m} M)] \subset \exp \frac{1}{n} M.$$

(d) Let $g \in \mathrm{AutD}$ and $n \in \mathbb{N}$ be given; then there exists some $m \in \mathbb{N}$ such that $g.(\exp \frac{1}{m} M).g^{-1} \subset \exp \frac{1}{n} M$.

Indeed, once $g \in \mathrm{AutD}$ has been fixed, by corollary 5.12 the adjoint mapping $g_{\#}^{-1}$: autD→autD of g^{-1} is an automorphism of the

Banach-Lie algebra autD. Therefore the set $g_{\#}^{-1}(\frac{1}{n}M)$ is a neighbourhood of 0 in autD and we may find some $m\in\mathbb{N}$ such that $\frac{1}{m}M\subset g_{\#}^{-1}(\frac{1}{n}M)$. Moreover, by proposition 5.13,

$$\exp(\frac{1}{m}M)\subset\exp[g_{\#}^{-1}(\frac{1}{n}M)] = g_{\#}^{-1}(\frac{1}{n}M) = g^{-1}.\exp(\frac{1}{n}M).g$$

so that $g.\exp(\frac{1}{m}M)g^{-1}\subset\exp\frac{1}{n}M$.

In order to show that $\mathcal{T}_a\geq\mathcal{T}$ it suffices to prove that every $\mathcal{T}$-neighbourhood of id_D contains a $\mathcal{T}_a$-neighbourhood of id_D. Now, the family of subsets of AutD given by

$$N(\varepsilon)=:\{g\in\mathrm{AutD};\ \|g-id_D\|_B<\varepsilon\}$$

for $\varepsilon>0$ is a fundamental system of $\mathcal{T}$-neighbourhoods of id_D. By lemma 6.45 the mapping exp: $M\to\mathrm{Hol}_\infty(B.E)$ is continuous at the origin; as

$$N(\varepsilon)=\{f\in\mathrm{Hol}_\infty(B,E);\ \|f-id_D\|_B<\varepsilon\}\cap\mathrm{AutD}$$

is a neighbourhood of id_D for the topology induced by $H_\infty(B,E)$ on AutD, there exists some $n\in\mathbb{N}$ such that $\exp(\frac{1}{n}M)\subset N(\varepsilon)$.

#

<u>6.53. DEFINITION</u>. *We refer to the topology introduced by theorem 6.52 on* AutD *as the "analytic topology" on* AutD.

By Aut_0D we denote the connected component of id_D in (AutD, $\mathcal{T}_a$).

<u>6.54. LEMMA.</u> *There is a neighbourhood M of the origin in* autD *such that* exp: $M\to\exp M$ *is a homeomorphism when both M and* $\exp M$ *are endowed with their respective topologies as subspaces of* autD *and* (AutD, $\mathcal{T}$).

Proof: Let M be as in theorem 6.50. By lemma 6.47, exp: $M\to\mathrm{Hol}_\infty(B,E)$ is a homeomorphism of M onto a neighbourhood $\exp M$ of id_D in $\mathrm{Hol}_\infty(B,E)$. Now it suffices to realize that exp takes its values not only in $\mathrm{Hol}_\infty(B,E)$ but in the subset $\mathrm{AutD}\subset\mathrm{Hol}_\infty(B,E)$ and that the topology induced by $\mathrm{Hol}_\infty(B,E)$ on AutD is precisely $\mathcal{T}$.

#

6.55. REMARK. Observe that exp: $M \to \exp M$ is a homeomorphism, too, for the topologies induced on M and $\exp M$ by autD and (AutD, T_a). Thus, in particular, T_a and T agree on the subset $\exp M$ of AutD, but from this fact we cannot conclude that T_a and T agree on the whole group AutD: whereas $\exp M$ is a neighbourhood of id_D for T_a, it may fail to be so for T.

§5.- The Banach-Lie group structure of AutD.

Now we are going to construct a real Banach-Lie group structure on AutD whose underlying topology is T_a. For this purpose, let M, M_1 and M_2 be as in theorem 6.50 and 6.52 so that

$$C(M_1, M_1) \subset M \qquad C(M_2, M_2) \subset M_1 \tag{6.26}$$

and exp: $M \to \exp M$ is a homeomorphism for the topologies induced by autD and (AutD, T_a). Let us denote by log: $\exp M \to M$ its inverse and write

$$F =: \{N;\ N \text{ open and } 0 \in N \subset M\}$$

Then, the family $\{\exp N;\ N \in F\}$ is a fundamental system of neighbourhoods of id_D for T_a.

6.56. THEOREM. *There is a unique real analytic Banach manifold structure on* (AutD, T_a) *for which the family*

$$\{(\exp M,\ \log_{|\exp M});\ N \in F\}$$

is a system of charts at the identity element id_D.

Proof: Let $g \in$ AutD be given. The left translation $Lg: f \to g \circ f$, $f \in$ AutD, is an automorphism of the topological group (AutD, T_a) whose inverse is $(Lg)^{-1} = Lg^{-1}$. Now we define a system of charts at $g \in$ AutD as the family of pairs

$$\{(g.\exp N\ ,\ \log Lg^{-1}{}_{|g.\exp N});\ N \in F\}$$

Clearly, condition M_1 of definition 6.1 is satisfied because

$$\bigcup_{\mathcal{N}\in F,\ g\in \mathrm{AutD}} g.\exp\mathcal{N}$$

is an open cover of AutD for the topology $\mathcal{T}_a$. Moreover, these local charts are analytically compatible in the real sense, i.e., they satisfy condition M_2, too. Indeed, assume that

$$f\in(g_1.\exp\mathcal{N}_1)\cap(g_2.\exp\mathcal{N}_2)$$

for some $g_1,g_2\in\mathrm{AutD}$ and $\mathcal{N}_1,\mathcal{N}_2\in F$. Then, there are $A_1\in\mathcal{N}_1$ and $A_1\in\mathcal{N}_1$ such that

$$g_1.\exp A_1= f= g_2.\exp A_2$$

Thus, for $A\in\mathcal{N}_1\cap\mathcal{N}_2$ we have

$$\log(g_1^{-1}g_2\exp A)= \log[(\exp A_1)f^{-1}f(\exp A_2)^{-1}\exp A]=$$

$$= \log[\exp A_1\exp(-A_2)\exp A]$$

Since $A_1,A_2\in\mathcal{M}_2$, from theorem 6.50 we derive

$$\exp A_1\exp(-A_2)= \exp C(A_1,-A_2)= \exp A_3$$

where, by (6.26), $A_3=: C(A_1,-A_2)$ is a fixed element of $\mathcal{M}_1$. Then, from theorem 6.50 we derive

$$\log[\exp A_1\exp(-A_2)\exp A]= \log(\exp A_3\exp A)=$$

$$= \log[\exp C(A_3,A)]= C(A_3,A)$$

Whence the transition homeomorphism corresponding to the charts $(g_1\exp\mathcal{N}_1,\ \log|_{g_1\exp\mathcal{N}_1})$ and $(g_2\exp\mathcal{N}_2,\ \log|_{g_2\exp\mathcal{N}_2})$ is given by $A\to C(A_3,A)$ which is a real analytic mapping.

#

<u>6.57. THEOREM</u>. *The manifold* (AutD, $\mathcal{T}_a$) *is a real Banach-Lie group whose Banach-Lie algebra is* autD.

Proof: By proposition 6.19 and 6.30 it suffices to prove the statements (a), (b) and (c) below.

(a) For every fixed $g \in \mathrm{Aut}D$, the mapping $Lg: g \to gf$ is real analytic on AutD.

Indeed, let $g \in \mathrm{Aut}D$ be fixed. Choose any $f \in \mathrm{Aut}D$ and any local chart $(f.\exp N, \log Lf^{-1}|_{f.\exp N})$ of $(\mathrm{Aut}D, \mathcal{T}_a)$ at f. Then $(gf\exp N, \log L(gf)^{-1}|_{gf\exp N})$ is a local chart at gf and the expression of Lg is these charts is given by

$$[\log L(gf)^{-1}]L(g)[\log Lf^{-1}]^{-1}A = \log[(fg)^{-1}gf\,\exp A] =$$

$$= \log \exp A = A$$

for $A \in N$. Thus, Lg is analytic.

(b) For every fixed $g \in \mathrm{Aut}D$, the mapping $Tg: \mathrm{Aut}D \to \mathrm{Aut}D$ given by $f \to gfg^{-1}$ is real analytic.

Indeed, let $g \in \mathrm{Aut}D$ be given. Choose any $f \in \mathrm{Aut}D$ and any local chart $(f.\exp N, \log Lf^{-1}|_{f.\exp N})$ at f. Since the adjoint mapping $g_\#: \mathrm{aut}D \to \mathrm{aut}D$ of g is continuous, there is some $N_1 \subset M$ such that $g_\#(N_1) \subset N$; therefore,

$$\log \exp g_\#(A) = g_\# A$$

for $A \in N_1$. Then $(f.\exp N_1, \log Lf^{-1}|_{f\exp N_1})$ and $(gfg^{-1}\exp N_1, \log L(gfg^{-1})^{-1}|_{gfg^{-1}\exp N_1})$ are local charts at f and gfg^{-1}. Moreover, the expression of Tg in these charts is given by

$$[\log L(gfg^{-1})^{-1}]Tg[\log Lf^{-1}]^{-1}A = \log[(gf^{-1}g^{-1})gf(\exp A)g^{-1}] =$$

$$= \log[g(\exp A)g^{-1}] = \log[g_\#(\exp A)] = \log[\exp g_\#(A)] = g_\# A$$

for $a \in N_1$. Thus Tg is analytic.

(c) The mapping $F: (\mathrm{Aut}D)\times(\mathrm{Aut}D) \to \mathrm{Aut}D$ given by

$(f,g)\to fg^{-1}$ is analytic in a neighbourhood of (id_D, id_D).

Indeed, let M_2 be as in (6.26). Then $(\exp M_2, \log_{|\exp M_2})$ is a chart of AutD at id_D and its "cartesian square" is a chart of AutD×AutD at (id_D, id_D). By (6.25) we have

$$\log \exp[C(A_1,A_2)] = C(A_1,A_2)$$

for all $A_1,A_2 \in M_2$. Then, it is easy to check that the expression of F in these charts is given by

$$(A_1,A_2)\to C(A_1,A_2)$$

which is real analytic.

#

§6.- The action of AutD on the domain D.

We endow the domain D with its underlying real analytic manifold structure and consider D×AutD as a product manifold. Then we define the *action* of AutD on D as the mapping ψ: D×AutD→D given by $(x,f)\to f(x)$.

6.58. THEOREM. *The mapping* ψ: $(x,f)\to f(x)$ *is real analytic on* D×AutD.

Proof: It suffices to prove its analyticity near the identity element id_D.

Now, let $x\in D$ be given and fix any ball $B\subset\subset D$ centered at x. Starting with this ball B we can construct a neighbourhood M of 0 in autD as we did in §4. Then $(B, id_B)\times(\exp M, \log_{|\exp M})$ is a local chart of D AutD at (x, id_D). Also, (D, id_D) is a chart of D at x. Thus, it suffices to show that the mapping

$$(6.27) \qquad (y,A)\to(\exp A)y,$$

which is the expression of ψ in these charts, is real analytic in $B\times M$.

By lemma 6.45, $(y,A)\to(y,\exp A)$ is real analytic in $B\times M$ with values in $B\times Hol_\infty(B,E)$. Obviously, the mapping $B\times Hol_\infty(B,E)\to E$ given by $(y,f)\to f(y)$ is separately holomorphic; therefore, by Hartog's theorem, it is holomorphic and, in particular, real analytic. But (6.27) is the composite of $(y,A)\to(y,\exp A)$ and $(y,f)\to f(y)$, whence the result follows. #

CHAPTER 7

BOUNDED CIRCULAR DOMAINS

In this chapter we shall study the group AutD for domains D
with some particular geometric properties.

§1.- The Lie algebra autD for circular domains.

     7.1. DEFINITION. *We say that a bounded domain* D *is
"circular" if* $0 \in D$ *and, for all* $x \in D$ *and all* $\lambda \in \mathbb{C}$ *with* $|\lambda| = 1$,
we have $\lambda x \in D$.

Throughout the whole chapter, D will stand for a bounded
circular domain.

     7.2. LEMMA. *Let* D *be a bounded circular domain. Then the
vector field* Z: $x \to ix$ *is complete in* D.

     Proof: For $t \in \mathbb{R}$, we define f^t: $x \to e^{it}x$. Since D is
circular, we have $f^t \in AutD$ and the mapping $t \to f^t$ is a
T-continuous one-parameter group. By theorem 4.5 its associated
vector field, which is obviously Z, is complete in D.

#

We call Z the *circular vector field* and it will play an
important role in the study of circular domains.

Since $0 \in D$, any (non necessarily complete) holomorphic vector
field X in D is uniquely determined by its Taylor series at 0.
We write P_n for the space of continuous n-homogeneous
polynomials P: $E \to E$, so that we have X= $\sum\limits_{n=0}^{\infty} P_n$ where
$P_n =: X_0^{(n} \in P_n$ for $n \in \mathbb{N}$.

Let $Z_{\#}=: [Z,\cdot]$ be the adjoint of Z; then we may apply $Z_{\#}$ to X and, by reiterating this operation and taking linear combinations, we obtain expressions of the form

$$P(Z_{\#})X= (a_0+a_1Z_{\#}+..+a_rZ_{\#}^r)X$$

where $P(\lambda)= a_0+a_1\lambda+..+a_r\lambda^r$ is a polynomial in the indeterminate λ.

<u>7.3. LEMMA.</u> *Let* $P(\lambda)\in\mathbb{C}[\lambda]$ *be any polynomial in* λ *and assume that* $X=\sum_{n=0}^{\infty} P_n$ *is a holomorphic vector field in* D. *Then we have*

$$P(Z_{\#})X=\sum_{n=0}^{\infty} P(ni-i)P_n$$

Proof: For the homogeneous components P_n of X we have

$$Z_{\#}P_n= [Z,P_n]= (n-1)iP_n$$

By reiterating this operation and taking linear combinations we obtain the result.

#

<u>7.4. LEMMA.</u> *Assume that* $X=\sum_{n=0}^{\infty} P_n$ *satisfies* $X\in\text{aut}D$. *Then we have* $P_n= 0$ *for all* $n\geq 3$.

Proof: Let $X\in\text{aut}D$ be given. Since autD is a real Lie algebra and $Z\in\text{aut}D$, we have $P(Z_{\#})X\in\text{aut}D$ for any polynomial with real coefficients $P(\lambda)\in\mathbb{R}[\lambda]$. By taking $P(\lambda)= \lambda^3+\lambda$ and applying lemma 7.3 we obtain

$$P(Z_{\#})X=\sum_{n=0}^{\infty} P(ni-i)P_n= P(-i)P_0+P(0)P_1+P(i)P_2+\sum_{n\geq 3} P(ni-i)P_n$$

But now we have $P(-i)= P(0)= P(i)= 0$, so that the Taylor series of $P(Z_{\#})X$ at 0 is

$$P(Z_{\#})X=\sum_{n\geq 3} P(ni-i)P_n$$

Thus, we have $[P(Z_{\#})X]_0^{(k}=0$ for $k=0,1$ and, by Cartan's uniqueness theorem, $P(Z_{\#})X=0$. However, $P(ni-i)\neq 0$ for all $n\geq 3$; therefore $P_n=0$ for $n\geq 3$.

7.5. DEFINITION. *For any bounded circular domain* D, *we set*

$\mathrm{aut}^0 D =: P_1 \cap (\mathrm{aut}D)$ $\qquad$ $\mathrm{aut}_0 D =: (P_0 \oplus P_2) \cap \mathrm{aut}D$

$E =: (\mathrm{aut}D)0 = \{X(0);\ X\in \mathrm{aut}D\}$ $\qquad$ $\mathrm{Aut}^0 D =: \{F\in \mathrm{Aut}D,\ F \text{ is linear}\}$.

7.6. PROPOSITION. *For bounded circular domains* D, *we have the topological direct sum decomposition*

$$\mathrm{aut}D = (\mathrm{aut}_0 D)\oplus(\mathrm{aut}^0 D) \tag{7.1}$$

Moreover, $\mathrm{aut}_0 D$ *is topologically isomorphic with* E_0 *(considered as a real linear subspace of* E*) by the mapping* $X\mapsto X(0)$ *and* $\mathrm{aut}^0 D$ *can viewed as the Lie algebra of* $\mathrm{Aut}^0 D$.

Proof: Let $X\in \mathrm{aut}D$ be given; by lemma 7.4 we have $X=P_0+P_1+P_2$ for some $P_k\in P_k$, $k=0,1,2$. Applying lemma 7.3 to the polynomial $P(\lambda)=\lambda^2$ and the vector field X we derive

$$P(Z_{\#})X=\sum_{n=0}^{2} P(ni-i)P_n=\sum_{n=0}^{2}(in-i)^2 P_n=-(P_0+P_2)$$

so that $P_0+P_2\in \mathrm{aut}D$ and, therefore, $P_1=X-(P_0+P_2)\in \mathrm{aut}D$. Clearly $P_0+P_2\in \mathrm{aut}_0 D$ and $P_1\in \mathrm{aut}^0 D$; thus autD admits the direct sum decomposition $\mathrm{aut}D=(\mathrm{aut}^0 D)\oplus(\mathrm{aut}_0 D)$. By lemma 5.11, the canonical projectors $Z_{\#}^2$ and $I-Z_{\#}^2$ are continuous.

Now, let $c\in E_0$ be given. Then, there exists a unique symmetric bilinear mapping $Q_c\in L(E\times E|E)$ such that the vector field $A: x\mapsto c-Q_c(x,x)$, $x\in D$, belongs to autD. Indeed, there is some $X=P_0+P_1+P_2\in \mathrm{aut}D$ with $c=X(0)=P_0$. Then we have $c+P_2\in \mathrm{aut}D$ and $Q(x,x)=:-P_2(x)$, $x\in D$, satisfies the requirements. If there is another Q'_c in the same conditions, from $c-Q_c\in \mathrm{aut}D$ and $c-Q'_c\in \mathrm{aut}D$ we get $Q_c-Q'_c=(c-Q'_c)-(c-Q_c)\in \mathrm{aut}D$; thus

$(Q_c - Q'_c)_0^{(k} = 0$ for $k = 0,1$ and, by Cartan's uniqueness theorem, $Q_c = Q'_c$.

Now, we show that E_0 is complete and that the mapping $E_0 \to \mathrm{aut}_0 D$ given by $c \to c - Q_c$ is a continuous surjective isomorphism of Banach spaces. Indeed, assume that $Q_c = 0$ for some $c \in E_0$. As $c - Q_c \in \mathrm{aut}_0 D$, we have $c \in \mathrm{aut} D$. From $Z \in \mathrm{aut} D$ we obtain $[Z, c] = ic \in \mathrm{aut} D$. Since autD is purely real, we have $c = 0$. Thus, $c \to c - Q_c$ is an isomorphism onto the image subspace which is obviously $\mathrm{aut}_0 D$. Now, let us take any ball $B \subset\subset D$ centered at $0 \in D$. By theorem 5.6, there are constants K_1, K_2 such that we have

$$K_1 \sum_{s=0}^{1} \| X_0^{(s} \| \leq \| X \|_B \leq K_2 \sum_{s=0}^{1} \| X_0^{(s} \|$$

for all $X \in \mathrm{aut} D$. Applying this to the vector field $X = c - Q_c \in \mathrm{aut}_0 D$ we obtain

$$K_1 \| c \| \leq \| c - Q_c \| \leq K_2 \| c \|$$

for all $c \in E_0$. Thus $c \to c - Q_c$ is a homeomorphism. Since we know that $\mathrm{aut}_0 D$ is closed in autD, E_0 is complete. As for the assertion concerning $\mathrm{aut}^0 D$, we can repeat the arguments of theorems 6.56 and 6.57 restricting ourselves to the group $\mathrm{Aut}^0 D$ instead of AutD.

In the course of the proof we have established the following

7.7. COROLLARY. E_0 *is a real subspace of* E *and, for each* $c \in E_0$, *there is a unique* $Q_c \in L(E \times E | E)$ *such that the vector field* $x \to c - Q_c(x,x)$, $x \in D$, *belongs to* autD.

7.8. DEFINITION. *We reserve the notation* Q_c *for the symmetric bilinear mapping described above.*

7.9. PROPOSITION. *For bounded circular domains* D, *we have*

$$[\mathrm{aut}^0 D, \mathrm{aut}^0 D] \subset \mathrm{aut}^0 D, \qquad [\mathrm{aut}^0 D, \mathrm{aut}_0 D] \subset \mathrm{aut}_0 D$$

$$[\mathrm{aut}^0 D, \mathrm{aut}^0 D] \subset \mathrm{aut}^0 D$$

Moreover,

(a) *For all* $L \in \mathrm{aut}^0 D$, $c \in E_0$ *and* $x \in E$, *it holds* $Lc \in E_0$ *and* $Q_{Lc}(x,x) = LQ_c(x,x) - 2Q_c(Lx,x)$.

(b) *For all* $c_1, c_2 \in E_0$, *we have* $Q_{c_1}(.,c_2) + Q_{c_2}(c_1,.) \in \mathrm{aut}^0 D$

(c) *For* $x \in E$ *and* $c_1, c_2 \in E_0$ *the following equality holds* $Q_{c_1}[Q_{c_2}(x,x),x] = Q_{c_2}[Q_{c_1}(x,x),x]$.

(d) *For all* $F \in \mathrm{Aut}^0 D$, $c \in E_0$ *and* $x \in E$, *we have* $Fc \in E_0$ *and* $Q_{Fc}(x,x) = FQ_c(F^{-1}x, F^{-1}x)$.

Proof: Let $L_1, L_2 \in \mathrm{aut}^0 D$ be given; then $[L_1, L_2]$ is linear so that $[\mathrm{aut}^0 D, \mathrm{aut}^0 D] \subset \mathrm{aut}^0 D$. In particular, $\mathrm{aut}^0 D$ is a Banach-Lie subalgebra of autD.

Let $L \in \mathrm{aut}^0 D$ and $A \in \mathrm{aut}_0 D$ be arbitrarily given; then we have $A(x) = c - Q_c(x,x)$, $x \in D$, where $c \in E_0$ and Q_c is a symmetric bilinear mapping $E \times E \to E$. An easy computation gives

$$[L,A]x = L(c) + 2Q_c(Lx,x) - L[Q_c(x,x], \qquad x \in D$$

Since the mapping $x \to 2Q_c(Lx,x) - L[Q_c(x,x)]$ is an element of P_2, *it follows* $[L,A] \in \mathrm{aut}_0 D$. Besides $Lc = [L,A]0 \in E_0$ and

$$Q_{Lc}(x,x) = LQ_c(x,x) - 2Q_c(Lx,x), \qquad x \in E$$

which proves (a).

Let $A_1, A_2 \in \mathrm{aut}_0 D$ be given and assume that

$$A_1(x) = c_1 - Q_{c_1}(x,x) \qquad\qquad A_2(x) = c_2 - Q_{c_2}(x,x)$$

where $c_1, c_2 \in E_0$ and Q_{c_1}, $Q_{c_2} \in \mathcal{L}(E \times E | E)$ are symmetric. It follows that

$$[A_1, A_2]x = 2\{Q_{c_2}(c_1,x) + Q_{c_1}(x,c_2)\} + 2\{Q_{c_2}(x, Q_{c_1}(x,x)) -$$

$$-Q_{c_1}(Q_{c_2}(x,x),x)\}$$

Since the mapping $x \to Q_{c_2}(x,Q_{c_1}(x,x)) - Q_{c_1}(Q_{c_2}(x,x),x)$, $x \in D$, is an element of P_3, by lemma 7.4, it must be identically null. Thus $[A_1,A_2] = Q_{c_1}(\cdot,c_2) + Q_{c_2}(c_1,\cdot) \in \mathrm{aut}^0 D$. This proves (b) and (c).

Finally, let $F \in \mathrm{Aut}^0 D$ and $a \in E_0$ be given. Then we have $F = \exp L$ for some $L \in \mathrm{aut}^0 D$, i.e., for $x \in D$, Fx is the value at $t = 1$ of the solution of the initial value problem

$$\text{(7.2)} \qquad \frac{d}{dt}\, y(t) = L[y(t)]\ , \qquad y(0) = x, \quad y(t) \in D$$

in the space E. By (a) we have $L(E_0) \subset E_0$; therefore, if the initial value is some $c \in E_0 \cap D$, (7.2) can be interpreted as an initial value problem in E_0. Since E_0 is complete and the solution of (7.2) is unique, we have $Fc = (\exp L)c \in E_0$ for all $c \in E_0 \cap D$. As F is linear, $Fc \in E_0$ for all $c \in E_0$. Moreover,

$$F_{\#}[c - Q_c(x,x)] = F[c - Q_c(F^{-1}x, F^{-1}x)] = Fc - FQ_c(F^{-1}x, F^{-1}x)$$

for all $x \in E$, so that

$$Q_{Fc}(x,x) = FQ_c(F^{-1}x, F^{-1}x)$$

which shows (e).

#

7.10. COROLLARY. *The subspace* E_0 *is invariant under the group* $\mathrm{Aut}^0 D$. *In particular,* E_0 *is a complex subspace of* E. *The mapping* $c \to Q_c$ *is conjugate linear and we have* $Q_c(c,\cdot) \in \mathrm{aut}^0 D$ *for all* $c \in E_0$.

Proof: We have $F =: \mathrm{id}_D \in \mathrm{Aut}^0 D$ because D is circular. Applying (e) we get $Q_{ic} = -iQ_c$. Then apply (b) with $c_1 = c_2 = c \in E_0$.

#

<u>7.11. LEMMA</u> . *We have* $E_0 = \{X(c);\ c \in E_0\ ,\ X \in \mathrm{aut}D\}$.

Proof: Let us set

$(\mathrm{aut}^0 D)E_0 =: \{L(c);\ c \in E_0,\ L \in \mathrm{aut}^0 D\}$, $(\mathrm{aut}_0 D)E_0 = \{A(c);\ c \in E_0, A \in \mathrm{aut}_0 D\}$

First we show that $(\mathrm{aut}^0 D)E_0 \subset E_0$. Indeed, let $c \in E_0$ be given and take any $L \in \mathrm{aut}^0 D$. Then $A =: c - Q_c \in \mathrm{aut}D$ so that $[L,A] \in \mathrm{aut}D$. But

$$[L,A]x = L(c) + L(Q_c(x,x)) + 2Q_c(x,L(x)), \quad x \in D$$

By evaluating $[L,A]$ at 0 we get $L(c) \in E_0$ and therefore $(\mathrm{aut}^0 D)E_0 \subset E_0$

Now we show that $(\mathrm{aut}_0 D)E_0 \subset E_0$. Indeed, let $c \in E_0$ be given. Then $A = c - Q_c \in \mathrm{aut}D$; since $Z \in \mathrm{aut}D$, we have $A_\#^2(Z) = [A,[A,Z]] \in \mathrm{aut}D$. But

$$[A,[A,Z]]x = 4iQ_c(x,c) \qquad x \in D$$

Since the mapping $x \mapsto 4iQ_c(x,c)$ is linear, we must have $[A,[A,Z]] \in \mathrm{aut}^0 D$ and by the previous step we get

$$Q_c(x,c) \in E_0 \tag{7.3}$$

for all $c \in E_0$ and $x \in E_0$. Interchanging the roles of x and c, by the symmetry of Q_c we get

$$Q_x(c,x) \in E_0 \tag{7.4}$$

for all $x, c \in E_0$. As $c \mapsto Q_c$ is linear,

$$Q_{x+c}(x,x+c) = Q_{x+c}(x+c,x) = Q_x(x+c,x) + Q_c(x+c,x) \in E_0$$

Thus, from (7.3) and (7.4) we derive $Q_c(x+c,x) \in E_0$ for all $x, c \in E_0$. Then, from

$$Q_c(x+c,x) = Q_c(x,x) + Q_c(c,x) \in E_0$$

and (7.3) we obtain $Q_c(x,x) \in E_0$ for all $x \in E_0$, so that

$(\mathrm{aut}_0 D)E_0 \subset E_0$.

Now, by proposition 7.6 it follows

$(\mathrm{aut}D)E_0 =: \{X(c);\ c\in E_0\ ,\ X\in \mathrm{aut}D\} \subset (\mathrm{aut}^0 D)E_0 + (\mathrm{aut}_0 D)E_0 \subset E_0$.

The converse inclusion is obvious.

#

§2.- The connected component of the identity in AutD.

Let $\mathcal{J}$ be any Hausdorff topological group and denote by W an open symmetric connected neighbourhood of the identity element e in $\mathcal{J}$. For $n\in\mathbb{N}$ we set

$$W^n =: \{g_1\circ g_2\circ\ldots\circ g_n;\ g_k\in K,\ \ k=1,2,..,n\}$$

From the general theory of topological groups, it is known that

$$\mathcal{H} =: \bigcup_{n\in N} W^n$$

is a closed normal subgroup of $\mathcal{J}$ and that $\mathcal{H}$ is the connected component of e in $\mathcal{J}$; thus, $\mathcal{H}$ does not depend on the choice of W. Since the mapping *exp* is a local homeomorphism at 0, in particular we get.

7.12. LEMMA. *Let M be the neighbourhood of* 0 *in* aut D *given by lemma 6.54. Then*

$$\mathrm{Aut}_0 D =: \{(\exp A_1)\circ\ldots\circ(\exp A_n);\ A_k\in M,\ \ k=1,..,n, n\in\mathbb{N}\}$$

is the connected component of id_D *in* $(\mathrm{Aut}D,\mathcal{T}_a)$. *Moreover,* $\mathrm{Aut}_0 D$ *is a closed normal subgroup of both* $(\mathrm{Aut}D,\mathcal{T}_a)$ *and* $(\mathrm{Aut}D,\mathcal{T})$.

7.13. DEFINITION. *Let* S *and* G *be subsets of* D *and* autD *respectively. We define the "orbit" of* S *by* G *by means of*

$$G(S) =: \{g(x);\ g\in G\ ,\ x\in S\}.$$

It is immediate to check that if G is a subgroup of AutD, then we have $\mathcal{G}\mathcal{G}(S) = \mathcal{G}(S)$.

Let E_0 be the subspace of E given by definition 7.5. Clearly $E_0 \cap D$ is a bounded open circular subset of the space E_0; however, $E_0 \cap D$ may fail to be connected

7.14. LEMMA. *If* D *is a bounded circular domain of* E, *then we have* $(\mathrm{Aut}_0 D)(E_0 \cap D) \subset E_0 \cap D$.

Proof: Let $g \in \mathrm{Aut}_0 D$ be given. By lemma 7.12 we have $g = (\exp A_1) \circ \ldots \circ (\exp A_n)$ for some $A_k \in M$, $k = 1,2,..,n$. Thus, it suffices to show that

$$(\exp A)(E_0 \cap D) \subset E_0 \cap D$$

for all $A \in M$.

Now, let $A \in M$ and $x \in E_0 \cap D$ be given. Let us consider the initial value problem

$$\frac{d}{dt}\, y(t) = A[y(t)], \qquad y(0) = x, \qquad y(t) \in D \tag{7.5}$$

in the space E. Its solution $y(t) = (\exp tA)x$ satisfies $y(t) \in D$ for all $t \in \mathbb{R}$. Since $x \in E_0$ and by proposition 7.6 E_0 is complete, the initial value problem (7.5) has a solution in E_0, too. As the solution is unique, we have $(\exp tA)x \in E_0$ for all $t \in \mathbb{R}$; thus, $(\exp A)(E_0 \cap D) \subset E_0 \cap D$.

#

7.15. LEMMA. *Let* D *be a bounded circular domain of* E. *Then, if* $E_0 \cap D$ *is connected (in particular, if* D *is balanced), we have*

$$(\mathrm{Aut}_0 D)|_{E_0 \cap D} \subset \mathrm{Aut}(E_0 \cap D)$$

where

$$(\mathrm{Aut}_0 D)|_{E_0 \cap D} =: \{g|_{E_0 \cap D};\ g \in \mathrm{Aut}_0 D\}$$

Proof: As $E_0 \cap D$ is assumed to be connected, $E_0 \cap D$ is a bounded circular domain of the space E_0 and it makes sense to speak of the group $\mathrm{Aut}(E_0 \cap D)$.

Let $g \in \mathrm{Aut}_0 D$ be given. Then g is a biholomorphic bijection of D onto D; therefore $g|_{E_0 \cap D}$ is a biholomorphic bijection of $E_0 \cap D$ onto its image $g(E_0 \cap D)$. From lemma 7.14,

$$g(E_0 \cap D) \subset E_0 \cap D \tag{7.6}$$

Thus, applying $g^{-1} \in \mathrm{Aut}_0 D$ to (7.6) and lemma 7.14 again, we obtain

$$E_0 \cap D \subset g^{-1}(E_0 \cap D) \subset E_0 \cap D$$

so that $g^{-1}(E_0 \cap D) = E_0 \cap D$. Then $g(E_0 \cap D) = E_0 \cap D$ and $g|_{E_0 \cap D} \in \mathrm{Aut}(E_0 \cap D)$.

Observe that, if D is balanced, then $E_0 \cap D$ is balanced too; hence it is connected and the lemma holds. #

<u>7.16. LEMMA</u>. *Let* D *be a bounded circular domain of* E. *Then, if* $E_0 \cap D$ *is connected (in particular, if* D *is balanced), the set* $(\mathrm{Aut}_0 D)0$ *is a neighbourhood of* 0 *in the space* E_0.

Proof: Consider the mapping $\phi: E_0 \to \mathrm{aut}_0 D$ given by $c \to c - Q_c$. Since M is a neighbourhood of 0 in $\mathrm{aut} D$, $M \cap \mathrm{aut} D_0$ is a neighbourhood of 0 in $\mathrm{aut}_0 D$ and, by proposition 7.6, $U =: \phi^{-1} M \cap \mathrm{aut}_0 D$ is a neighbourhood of 0 in E_0. Now, consider the composite J of the mappings

$$c \to c - Q_c \to \exp(c - Q_c) \to [\exp(c - Q_c)]0$$

By lemma 6.32 we have $J(c) = \sum_{n=0}^{\infty} \frac{1}{n!} (\hat{A}_c^n \mathrm{id}_D)0$ for $c \in U$, where we have put $A_c =: c - Q_c$. Moreover, we have $(\hat{A}_c^n \mathrm{id}_D) = 0$ for all $n \neq 1$ and $(\hat{A}_c^1 \mathrm{id}_D)0 = c$, so that $J(c) = c$ for $c \in U$ and $J(U)$ is a neighbourhood of 0. Thus, by lemma 7.12.

$$0 \in U = J(U) = \{(\exp A_c)0;\ c \in U\} \subset \{\exp A)0;\ A \in M\} \subset (\mathrm{Aut}_0 D)0$$

and, by lemmas 7.15 and 7.14,

$$(\mathrm{Aut}_0 D)0 \subset \mathrm{Aut}(E_0 \cap D) \subset E_0 \cap D$$

whence the result follows.

#

7.17. PROPOSITION. *Let* D *be a (non necessarily circular) bounded domain in* E *and let* $\mathcal{J}$ *be a subgroup of* AutD *such that, for some* $x \in D$, *the orbit* $\mathcal{J}(x)$ *of* x *by* $\mathcal{J}$ *is a neighbourhood of* x. *Then* $\mathcal{J}(x) = D$ *and the subgroup* $\mathcal{J}$ *acts transitively on* D.

Proof: First we show that $\mathcal{J}(x)$ is an open subset of D. Let $y \in \mathcal{J}(x)$ be given. Then we have $g(x) = y$ for some $g \in \mathcal{J}$. As $\mathcal{J}(x)$ is assumed to be a neighbourhood of x, there exists some open subset $W \subset D$ such that $x \in W \subset \mathcal{J}(x)$ and, applying g we obtain

$$y = g(x) \in g(W) \subset g\mathcal{J}(x) = \mathcal{J}(x)$$

Since g is a homeomorphism, g(W) is open; thus, by the arbitrariness of y, $\mathcal{J}(x)$ is open.

Now we show that $\mathcal{J}(x)$ is a closed subset of D. Let $y \in D$ be any point of the closure $\overline{\mathcal{J}(x)}$ of $\mathcal{J}(x)$ in D. Then, there is a sequence $(y_n)_{n \in N} \subset \mathcal{J}(x)$ such that $y_n \to y$. Therefore, we have $y_n = g_n(x)$ for some $g_n \in \mathcal{J}$ and $n \in \mathbb{N}$. Let d_D be Carathéodory distance in D. Since we have assumed that $\mathcal{J}(x)$ is a neighbourhood of x and, by corollary 3.14, d_D induces the norm topology on D,

$$\{z \in D;\ d_D(z,x) < \varepsilon\} \subset \mathcal{J}(x) \tag{7.7}$$

for some $\varepsilon > 0$. Moreover, as $y_n \to y$ it follows that

$$d_D(y_n,t) = d_D(g_n x, y) < \varepsilon \tag{7.8}$$

for all $n \geq n_0$. Since d_D is $\mathcal{J}$-invariant, from (7.8) we obtain $d_D(x, g_{n_0}^{-1} y) < \varepsilon$ so that, by (7.7), $g_{n_0}^{-1} y \in \mathcal{J}(x)$ and therefore $y \in g_{n_0} \mathcal{J}(x) = \mathcal{J}(x)$. Thus $\mathcal{J}(x)$ is closed in D.

Since D is connected, we have $\mathcal{J}(x) = D$ and D is homogeneous under the action of J.

#

7.18. COROLLARY. *Let* D *be a bounded circular domain of* E. *Then, if* $E_0 \cap D$ *is connected (in particular, if* D *is balanced) we have* $(\mathrm{Aut}_0 D)0 = E_0 \cap D$.

Proof: Consider the Banach space E_0, the bounded domain $E_0 \cap D$ and the point $0 \in E_0 \cap D$. By lemma 7.15, $\mathrm{Aut}_0 D$ is a subgroup of $\mathrm{Aut}(E_0 \cap D)$; by lemma 7.16 $(\mathrm{Aut}_0 D)0$ is a neighbourhood of 0 in $E_0 \cap D$. Then, proposition 7.17 gives the result.

#

§3.- Study of the orbit (AutD)0 of the origin.

In order to make a deeper study of the orbit (AutD)0 of the origin we recall some properties of analytic sets.

7.19. DEFINITION. *A subset* Ω *of domain* D *is said to be complex-analytic in* D *if, for every point* $x \in D$, *there is a neighbourhood* U *of* x *and there is a set* $F_x \subset \mathrm{Hol}(U,\mathbb{C})$ *of holomorphic functions* $f: U \to \mathbb{C}$ *such that we have*

$$\Omega \cap U = \{y \in U;\ f(y) = 0 \quad \forall f \in F_x\}$$

Roughly speaking, a subset Ω of D is analytic in D if, and only if, Ω can be locally represented as the "joint kernel" of a set of holomorphic functions. For a study of the elementary properties of analytic sets see for example |45| p. 50.

7.20. DEFINITION. *Let* D *be a bounded circular domain in* E. *We denote by* Ω *the subset of* D *consisting of the points* $x \in D$ *for which* $(\mathrm{Aut}_0 D)x$ *is a complex-analytic closed set in* D.

7.21. LEMMA. *Let* D *be a bounded circular domain in* E. *Then, if* $E_0 \cap D$ *is conneted (thus, in particular, when* D *is balanced), we have* $\Omega \neq \phi$ *and* $(\mathrm{AutD})\Omega = \Omega$.

Proof: Since $E_0 \cap D$ is assumed to be connected, by corollary 7.18 we have $(\mathrm{Aut}_0 D)0 = E_0 \cap D$. Since E_0 is a closed complex subspace of E, by the Hanh-Banach separation theorem, $E_0 \cap D$ is a complex-analytic closed set in D; thus se have $0 \in \Omega$ and $\Omega \neq \phi$.

Let $x \in (\mathrm{aut}D)\Omega$; then there are $y \in \Omega$ and $g \in \mathrm{Aut}D$ such that $gy = x$. Since by lemma 7.12 $\mathrm{Aut}_0 D$ is a normal subgroup of AutD, we have $(\mathrm{aut}_0 D)x = (\mathrm{Aut}_0 D)gy = g(\mathrm{aut}_0 D)y$. But $(\mathrm{Aut}_0 D)y$ is a complex-analytic closed set in D and these properties are preserved by $g \in \mathrm{Aut}D$; thus, $x \in \Omega$ and $(\mathrm{Aut}D)\Omega \subset \Omega$. The oposite inclusion is obvious.

#

7.22. LEMMA. *Let* D *be a bounded circular domain for which* $E_0 \cap D$ *is connected. Then, for all* $x \in \Omega$, *we have*

$$\{\lambda x;\ \lambda \in \mathbb{C},\ |\lambda| \leqslant 1\} \subset (\mathrm{Aut}_0 D)x$$

Proof: If $x=0$, then the assertion is trivial. Let $x \in \Omega$ be given with $x \neq 0$ and put $V =: \{\lambda \in \mathbb{C};\ |\lambda| < \|x\|^{-1}\}$. Since $(\mathrm{Aut}_0 D)x$ is a complex-analytic closed set in D,

$$W =: \{\mu \in \mathbb{C};\ \mu x \in (\mathrm{Aut}_0 D)x\}$$

is a complex-analytic closed subset of V. As D is circular, the circular group $f^t(y) =: e^{it}y$, $t \in \mathbb{R}$, $y \in D$, is contained in $\mathrm{Aut}_0 D$; therefore

$$e^{it}x \in e^{it}(\mathrm{Aut}_0 D)x = (\mathrm{Aut}_0 D)x$$

for all $t \in \mathbb{R}$, so that $\{e^{it};\ t \in \mathbb{R}\} \subset W$. Since W is an analytic subset of V, $V \setminus W$ is connected (|43| proposition 1 page 50). Then the unit disc of $\mathbb{C}$ is contained in W, whence the result follows.

#

7.23. PROPOSITION. *Let* D *be a bounded circular domain in* E. *Then, if* $E_0 \cap D$ *is connected (thus, in particular, if* D *is balanced), we have*

$$(\mathrm{Aut}D)0 = (\mathrm{Aut}_0 D)0 = \Omega = E_0 \cap D$$

Proof: First we show that $\Omega \subset (\mathrm{Aut}_0 D)0$. Let $x \in \Omega$ be given. By lemma 7.22 we have $0 \in (\mathrm{Aut}_0)x$; thus, $gx = 0$ for some $g \in \mathrm{Aut}_0 D$. Then we have $x = g^{-1}0 \in g^{-1}(\mathrm{Aut}_0 D)0 = (\mathrm{Aut}_0)0$ and $\Omega \subset (\mathrm{Aut}_0 D)0$.

Now we show that $(\mathrm{Aut}D)\Omega \subset \Omega$. Obviously $0 \in \Omega$; by lemma 7.21, Ω is invariant under AutD, so that applying AutD to the relation $0 \in \Omega$ we get $(\mathrm{Aut}D)0 \subset (\mathrm{Aut}D)\Omega \subset \Omega$.

Thus, we have $\Omega \subset (\mathrm{Aut}_0 D)0 \subset (\mathrm{Aut}D)0 \subset \Omega$ and corollary 7.18 completes the proof.

#

7.24. COROLLARY. *Let* D *be a bounded circular domain in* E. *Then, if* $E_0 \cap D$ *is connected (in particular, if D is balanced), the orbit* $(\mathrm{Aut}D)0$ *is balanced.*

Proof: Let $x \in (\mathrm{Aut}D)0$ and $\lambda \in \mathbb{C}$, $|\lambda| \leq 1$, be given. By proposition 7.23 we have $x \in \Omega$; then by lemma 7.22 we have $\lambda x \in (\mathrm{Aut}_0 D)x \subset (\mathrm{Aut}_0 D)(\mathrm{Aut}D)0 = (\mathrm{Aut}D)0$.

#

§4.- The decomposition $\mathrm{Aut}D = (\mathrm{Aut}^0 D)(\mathrm{Aut}_0 D)$.

7.25. THEOREM. *Let* $D \subset E$ *and* $\tilde{D} \subset \tilde{E}$ *be bounded circular domains in the Banach spaces* E *and* $\tilde{E}$ *respectively, and assume that* $f\colon D \to \tilde{D}$ *is an analytic isomorphism of* D *onto* $\tilde{D}$ *such that* $f(0) = 0$. *Then, there is a surjective continuous linear map* $F \in \mathcal{L}(E, \tilde{E})$ *such that* $F_{|D} = f$.

Proof: For $t \in \mathbb{R}$ we define $g^t\colon E \to E$ and $\tilde{g}^t\colon \tilde{E} \to \tilde{E}$ by means of $g^t(x) =: e^{it}x$ and $\tilde{g}^t(y) =: e^{it}y$. Obvioysly $g^t \in \mathrm{Aut}D$, $\tilde{g}^t \in \mathrm{Aut}\tilde{D}$ and, as $f\colon D \to \tilde{D}$ is a surjective isomorphism, the mapping $h =: g^{-t}f^{-1}\tilde{g}^t f$ satisfies $h \in \mathrm{Aut}D$ and $h(0) = 0$. From the chain rule and the fact that we have $Lg^t = \tilde{g}^t L$ for all $L \in \mathcal{L}(E, \tilde{E})$, we derive

$$h_0^{(1} = (g^{-t})_0^{(1}(f^{-1})_0^{(1}(\tilde{g}^t)_0^{(1} = g^{-t}(f^{-1})_0^{(1}f_0^{(1}g^t = \mathrm{id}$$

so that, by Cartan's uniqueness theorem, we get $h = \mathrm{id}_D$ and

$\tilde{g}^t f = f g^t$, i.e.

$$e^{it} f(x) = f(e^{it} x)$$

for all $t \in \mathbb{R}$ and $x \in D$. By developing both $\tilde{g}^t f$ and $f g^t$ into their Taylor series at 0 we obtain

$$e^{it} \sum_{n \geq 1} f_0^{(n}(x,..,x) = \sum_{n \geq 1} e^{nit} f_0^{(n}(x,..,x), \quad x \in D$$

since we have assumed that $f(0) = 0$. Thus, by the uniqueness of the Taylor series,

$$[e^{(n-1)it} - 1] f_0^{(n}(x,..,x) = 0$$

for $t \in \mathbb{R}$, $n \geq 2$, and $x \in D$; therefore $f_0^{(n} = 0$. It follows that $f(x) = f_0^{(1}(x)$ for $x \in D$ and $f = F_{|D}$ where $F =: f_0^{(1} \in L(E, \tilde{E})$ is surjective.

#

<u>7.26. COROLLARY</u>. *Let* D *and* $\tilde{D}$ *be the open unit balls of* E *and* $\tilde{E}$ *and assume that* $f: D \to \tilde{D}$ *is a holomorphic map of* D *onto* $\tilde{D}$ *such that* $f(0) = 0$. *Then* f *is an isomorphism if and only if we have* $f = F_{|D}$ *for some surjective linear isometry* $F: E \to \tilde{E}$.

Proof: Assume that f is an isomorphism. By theorem 7.25 we have $f = F_{|D}$ for some surjective $F \in L(E, \tilde{E})$. Since $f(D) = F(D) = \tilde{D}$, F is an isometry. The converse is clear.

#

<u>7.27. DEFINITION</u>: *Let* D *be a bounded circular domain in* E. *We define the "isotropy subgroup" of the origin*, Isot D, *by means of*

$$\text{Isot } D =: \{F \in \text{Aut}D;\ F(0) = 0\}$$

Obviously Isot D is a closed subgroup of AutD for the topologies T and T_a. Moreover, from corollary 7.26 we immediately obtain that Isot D= $\text{Aut}^0 D$. The circular subgroup Z satisfies $Z \subset (\text{Aut}^0 D) \cap (\text{Aut}_0 D)$.

7.28. THEOREM. *Let* D *be a bounded circular domain in* E. *Then, if* $E_0 \cap D$ *is connected (in particular, if* D *is balanced), we have*

$$\mathrm{Aut}D = (\mathrm{Aut}^0 D)(\mathrm{Aut}_0 D) = (\mathrm{Aut}_0 D)(\mathrm{Aut}^0 D)$$

Proof: Let $f \in \mathrm{Aut}D$ be given. From proposition 7.23 we derive $f(0) \in (\mathrm{Aut}D)0 = (\mathrm{Aut}_0 D)0$; thus we have $f(0) = g(0)$ for some $g \in \mathrm{Aut}_0 D$. Then $h =: g^{-1} f \in \mathrm{Aut}D$ and $h(0) = g^{-1} f(0) = 0$, whence by theorem 7.25 we obtain $h = F_{|D}$ for some surjective $F \in L(E,E)$, so that $h \in \mathrm{Aut}^0 D$. Thus we have $f = gh$ with $g \in \mathrm{Aut}_0 D$ and $h \in \mathrm{Aut}^0 D$. The other equality comes from the fact that $\mathrm{Aut}_0 D$ is a normal subgroup of AutD.

Remark that the factorization $f = g.h$, with g and h in the above conditions, is not unique as the circular subgroup satisfies $Z \subset (\mathrm{Aut}_0 D) \cap (\mathrm{Aut}^0 D)$.

#

§5.- Holomorphic and isometric linear equivalence of Banach spaces.

Let E and $\tilde{E}$ be complex Banach spaces and D, $\tilde{D}$ their respective open unit balls.

7.29. DEFINITION. (a) *We say that* E *and* $\tilde{E}$ *are "holomorphically equivalent" if there is some surjective analytic isomorphism* $f: D \to \tilde{D}$.

(b) *We say that* E *and* $\tilde{E}$ *are "isometrically linearly equivalent" if there exists some surjective linear isometry* $L: E \to \tilde{E}$.

7.30. THEOREM. E *and* $\tilde{E}$ *are isomorphically equivalent if and only if they isometrically linearly equivalent.*

Proof: The "if part" is obvious. Thus, let $f: \tilde{D} \to D$ be any surjective holomorphic isomorphism. Then, the mapping $f_\#: \mathrm{Aut}D \to \mathrm{Aut}\tilde{D}$ given by $g \mapsto f^{-1} g f$ is a surjective isomorphism of these groups, so that we have $\mathrm{Aut}\tilde{D} = f^{-1}(\mathrm{Aut}D) f$ and, therefore,

$$f(\mathrm{Aut}\tilde{D}) = (\mathrm{Aut}D)f \quad , \quad f^{-1}(\mathrm{Aut}D) = (\mathrm{Aut}\tilde{D})f^{-1}$$

Let us denote by E_0 and $\tilde{E}_0$ the Banach spaces associated with E and $\tilde{E}$ by definition 7.5, so that

$$(7.9) \qquad (\mathrm{Aut}D)0 = E_0 \cap D \quad , \quad (\mathrm{Aut}\tilde{D})0 = E_0 \cap \tilde{D}$$

by proposition 7.23. We claim that

$$(7.10) \qquad (\mathrm{Aut}D)0 = f(\mathrm{Aut}\tilde{D})0, \qquad (\mathrm{Aut}\tilde{D})0 = f^{-1}(\mathrm{Aut}D)0$$

Indeed, from $f^{-1}(\mathrm{Aut}D)f = \mathrm{Aut}\tilde{D}$ we get that

$$[f^{-1}(\mathrm{Aut}D)f]0 = (\mathrm{Aut}\tilde{D})0 = \tilde{E}_0 \cap \tilde{D}$$

is a complex-analytic closed set in $\tilde{D}$. Let us put $\xi =: f(0)$; then $f^{-1}(\mathrm{Aut}D)\xi$ is a complex analytic closed set in $\tilde{D}$ and, applying f we obtain that $(\mathrm{Aut}D)\xi$ is a complex analytic closed subset of D. Therefore, by proposition 7.23 we have $\xi \in E_0 \cap D$. Then, as the orbit $E_0 \cap D$ of 0 is AutD-invariant,

$$\tilde{E}_0 \cap \tilde{D} = (\mathrm{Aut}\tilde{D})0 = [f^{-1}(\mathrm{Aut}D)f]0 =$$

$$= f^{-1}(\mathrm{Aut}D)\xi \subset f^{-1}(\mathrm{Aut}D)(E_0 \cap D) \subset f^{-1}(E_0 \cap D)$$

so that, applying f we obtain

$$f(\tilde{E}_0 \cap \tilde{D}) \subset E_0 \cap D$$

In a similar manner we get

$$E_0 \cap D \subset f(\tilde{E}_0 \cap \tilde{D})$$

Thus we have $E_0 \cap D = f(\tilde{E}_0 \cap \tilde{D})$, which is equivalent to (7.10). From the second of these formulas we obtain $f^{-1}(0) \in (\mathrm{Aut}\tilde{D})0$; thus we have $f^{-1}(0) = \tilde{g}(0)$ for some $\tilde{g} \in \mathrm{Aut}\tilde{D}$. Then $h =: f\tilde{g}$ is an analytic isomorphism of D onto D with $h(0) = f\tilde{g}(0) = 0$ whence the result follows by corollary 7.26.

#

§6.- The group of surjective linear isometries of a Banach space.

Let E be a complex Banach space with unit ball D=: B(E). The group $\mathrm{Aut}^0 D$ of all surjective linear isometries of E turns out to be a subgroup of both AutD and $GL(\mathcal{L}(E))$, the linear group of E. We obtain some properties of $\mathrm{Aut}^0 D$ by looking at it as a subgroup of these two groups.

Let $\mathrm{aut}D = \mathrm{aut}^0 D \oplus \mathrm{aut}_0 D$ be the decomposition of the Lie algebra autD given by proposition 7.6. For $\alpha \in GL(\mathcal{L}(E))$, let $\alpha_\#$ be the adjoint of α (cf. definition 4.26).

7.31. PROPOSITION. *Assume that* D *is homogeneous. Then we have the following characterization of* $\mathrm{Aut}^0 D$ *as a subgroup of* $GL(\mathcal{L}(E))$:

$$\mathrm{Aut}^0 D = \{\alpha \in GL(\mathcal{L}(E));\ \alpha_\#(\mathrm{aut}_0 D) \subset \mathrm{aut}_0 D\}$$

Proof: Let $\alpha \in GL(\mathcal{L}(E))$ be such that $\alpha \in \mathrm{Aut}^0 D$. It is an immediate consequence of the proof of proposition 7.9 (d) that $\mathrm{aut}_0 D$ is $\alpha_\#$-invariant (even if D is not homogeneous).

In order to prove the converse statement, we show first that the relation

$$\alpha_\#(\mathrm{aut}_0 D) \subset \mathrm{aut}_0 D$$

implies

$$\alpha_\#(\mathrm{Aut}_0 D) \subset \mathrm{Aut}_0 D$$

Indeed, let $f \in \mathrm{Aut}_0 D$ be given; by lemma 7.12 we can find $A_k \in \mathrm{aut}_0 D$, k= 1,2,...,n, such that

$$f = (\exp A_1) \circ \ldots \circ (\exp A_n)$$

Therefore, by proposition 5.13 we have

$$\alpha_\# f = \alpha f \alpha^{-1} = [\alpha(\exp A_1)\alpha^{-1}] \circ \ldots \circ [\alpha(\exp A_n)\alpha^{-1}] =$$

$$= \alpha_\#(\exp A_1) \circ \ldots \circ \alpha_\#(\exp A_n) = \exp(\alpha_\# A_1) \circ \ldots \circ \exp(\alpha_\# A_n)$$

From the assumptuion $\alpha_{\#}(\mathrm{aut}_0 D) \subset \mathrm{aut}_0 D$ we get $\alpha_{\#} A_k \in \mathrm{aut}_0 D$ for $k = 1,2,..,n$ so that, again by lemma 7.12, $\exp(\alpha_{\#} A_k) \in \mathrm{Aut}_0 D$ and, finally

$$\alpha_{\#} f = \exp(\alpha_{\#} A_1) \circ \ldots \circ \exp(\alpha_{\#} A_0) \in \mathrm{Aut}_0 D.$$

Now, let $\alpha \in GL(\mathcal{L}(E))$ be such that $\alpha_{\#}(\mathrm{aut}_0 D) \subset \mathrm{aut}_0 D$. Since D is assumed to be homogeneous, by proposition 7.23 it follows that

$$D = (\mathrm{Aut} D)0 = (\mathrm{Aut}_0 D)0$$

so that

$$\alpha(D) = [\alpha(\mathrm{Aut}_0 D)]0 \subset (\mathrm{Aut}_0 D)0 = D$$

A similar argument with $\alpha^{-1} \in GL(\mathcal{L}(E))$ gives $\alpha(D) = D$. Thus, α is a surjective linear isometry, i.e., $\alpha \in \mathrm{Aut}^0 D$.

#

<u>7.32. COROLLARY</u>. *If the unit ball* D *of* E *is homogeneous, then* $\mathrm{Aut}^0 D$ *is a real algebraic subgroup of degree* 2 *of* $GL(\mathcal{L}(E))$. *In particular,* $\mathrm{Aut}^0 D$ *is a Banach-Lie group for the topology of uniform convergence on* D.

Proof: By proposition 7.31 we have

$$\mathrm{Aut}^0 D = \{\alpha \in GL(\mathcal{L}(E));\ \alpha_{\#}(\mathrm{aut}_0 D) \subset \mathrm{aut}\ D_0\}$$

Let $A \in \mathrm{aut}_0 D$ be given. By proposition 7.6

$$A(x) = c - Q_c(x,x), \qquad x \in D$$

for some $c \in E$. Then $\alpha_{\#} A$ has the expression (cf. definition 4.26)

$$(\alpha_{\#} A)x = \alpha(c) - \alpha Q_c(\alpha^{-1} x,\ \alpha^{-1} x), \quad x \in D$$

Thus, again by proposition 7.31, $\alpha_{\#} A$ belongs to $\mathrm{aut}_0 D$ if, and only if.

$$\alpha Q_c(\alpha^{-1} x, \alpha^{-1} x) = Q_{\alpha(c)}(x,x) \qquad \forall x \in D$$

which is equivalent to

(*) $$\alpha Q_{\alpha^{-1}(c)}(x,x) = Q_c(\alpha x,\alpha x) \qquad \forall x\in D$$

Now, for fixed $c\in E$ and $x\in D$, the mappings $\mathcal{L}(E)\times\mathcal{L}(E)\to E$ given respectively by

$$q_{c,x}: (\alpha,\beta)\to Q_c(\alpha x,\alpha x) \qquad \ell_{c,x}:(\alpha,\beta)\to\alpha Q_{\beta(c)}(x,x)$$

are obviously continuous homogeneous polynomials of degree 2 and 1, and (*) can be reformulated as

$$q_{c,x}(\alpha,\alpha^{-1})-\ell_{c,x}(\alpha,\alpha^{-1}) = 0$$

with $c\in E$, $x\in D$ and $\alpha\in GL(\mathcal{L}(E))$. This a set of equations defining $\mathrm{Aut}^0 D$ as a real algebraic subgroup of degree 2 of $GL(\mathcal{L}(E))$. Then, theorem 6.40 completes the proof.

#

7.33. EXERCISE. Assume that the unit ball D of E is homogeneous. Show that, on the group $\mathrm{Aut}^0 D$, the analytic topology $\mathcal{T}_a$, the topology of local uniform convergence $\mathcal{T}$ and the topology $\mathcal{T}_u$ of uniform convergence over D coincide.

§7.- Boundary behaviour and extension theorems.

We recall that if D is a bounded circular domain, then any $A\in\mathrm{aut}D$ admits a unique representation of the form

$$A(x) = c+L(x)-Q_c(x,x), \qquad x\in E$$

where $c\in E_0$, $L\in\mathcal{L}(E)$ and $Q_c: E\to E$ is a continuous symmetric bilinear mapping. In particular, A is and entire mapping.

7.34. DEFINITION. *Let* $A\in\mathrm{aut}D$ *be given; then we set*

$$\rho =: \sup\{1,\ \|x\|\ ;\ x\in D\} \qquad C =: \|L\|+2\rho\|Q_c\|$$

Thus the numbers ρ *and* C *depend on the domain* D *and on the vector field* $A\in\mathrm{aut}D$.

For any fixed $\xi\in E$ and $A\in\mathrm{aut}D$, we can consider the initial value problem

$$\frac{dy}{dt} = A[y(t)], \qquad y(0) = \xi \tag{7.11}$$

whose maximal solution $\psi_\xi(t)$ is defined in a domain dom ψ_ξ of $\mathbb{R}$. Of course, if $\xi \in D$ then we have $\psi_\xi(t) = (\exp tA)\xi$ and $\mathrm{dom}\psi_\xi = \mathbb{R}$.

7.35. LEMMA. *Let* $A \in \mathrm{aut}D$ *and* $\xi \in E$ *be given. Then:*

(a) $\mathrm{dom}\psi_\xi \supset (-C^{-1}\log[1+\mathrm{dist}(\xi,0)^{-1}],\ C^{-1}\log[1+\mathrm{dist}(\xi,D)^{-1}])$

(b) *For any* $x \in D$ *and any* $t \in \mathbb{R}$ *with* $|t| < C^{-1}\log(1+\|\xi-x\|^{-1})$ *we have*

$$\|\psi_\xi(t)-\psi_x(t)\| < [(1+\|\xi-x\|^{-1})e^{-C|t|}-1]^{-1}$$

Proof: For $t \in \mathrm{dom}\psi_\xi$ we put $\delta(t) =: \|\psi_\xi(t)-\psi_x(t)\|$. We know that $\psi_\xi(0) = \xi$ and $\frac{d}{dt}\psi_\xi(t) = A[\psi_\xi(t)]$. Moreover, from the theory of ordinary differential equations, we know that the function $t \to \delta(t)$ is absolutely continuous in $\mathrm{dom}\psi_\xi$; thus, by Lebesgue's theorem, δ is derivable almost everywhere in $\mathrm{dom}\psi_\xi$. Besides, δ has right and left derivatives at any point of $\mathrm{dom}\psi_\xi$. From

$$\begin{aligned}\delta(t_1)-\delta(t_2) &= \|\psi_\xi(t_1)-\psi_x(t_1)\| - \|\psi_\xi(t_2)-\psi_x(t_2)\| \leqslant \\ &\leqslant \|[\psi_\xi(t_1)-\psi_\xi(t_2)]-[\psi_x(t_1)-\psi_x(t_2)]\|\end{aligned}$$

we obtain

$$\begin{aligned}\frac{d^+}{dt}\delta(t) &=: \lim_{\tau\downarrow 0}\frac{1}{\tau}[\delta(t+\tau)-\delta(t)] \leqslant \\ &\leqslant \overline{\lim_{\tau\downarrow 0}}\frac{1}{\tau}\|[\psi_\xi(t+\tau)-\psi_\xi(t)]-[\psi_x(t+\tau)-\psi_x(t)]\| \leqslant \\ &\leqslant \|\overline{\lim_{\tau\downarrow 0}}\{\frac{1}{\tau}[\psi_\xi(t+\tau)-\psi_\xi(t)]-\frac{1}{\tau}[\psi_x(t+\tau)-\psi_x(t)]\}\| = \\ &= \|\psi'_\xi(t)-\psi'_x(t)\| = \|A[\psi_\xi(t)]-A[\psi_x(t)]\| \leqslant \\ &\leqslant \|L[\psi_\xi(t)-\psi_x(t)]\| + \|Q_c[\psi_\xi(t),\psi_\xi(t)]-Q_c[\psi_x(t),\psi_x(t)]\|\end{aligned}$$

$$\leq \delta(t)\,\|L\| + \delta(t)\left[2\rho+\delta(t)\right]\,\|Q_c\| \leq C\left[\delta(t)+\delta^2(t)\right]$$

Thus, we have the relation

$$\frac{d}{dt}\,\delta(t) \leq C\left[\delta(t)+\delta^2(t)\right] \tag{7.12}$$

which is valid almost everywhere in $\mathrm{dom}\psi_\xi$. For $x \neq \xi$ we have $\delta(t)+\delta^2(t) \neq 0$ whatever is $t\in\mathrm{dom}\psi_\xi$, and from (7.12) we get

$$\frac{d}{dt}\log\frac{\delta(t)}{1+\delta(t)} \leq C$$

almost everywhere in $\mathrm{dom}\psi_\xi$, whence we easily derive

$$\delta(t) \leq e^{Ct}\left[(1+\delta(0)^{-1})e^{-Ct}-1\right]^{-1} \tag{7.13}$$

for all $t\in\mathrm{dom}\psi_\xi$.

On the other hand,

$$\delta(t) = \|\psi_\xi(t)-\psi_x(t)\| \geq \|\psi_\xi(t)\| - \|\psi_x(t)\| \geq \|\psi_\xi(t)\| - \rho$$

for $t\in\mathrm{dom}\psi_\xi$. Let us set

$$t^* =: \sup\{t\in\mathbb{R};\ t\in\mathrm{dom}\psi_\xi\}$$

From the extension theorems for ordinary differential equations we get

$$\overline{\lim_{t\to t^*}}\ \|\psi_\xi(t)\| = \infty$$

so that we have $\lim_{t\to t^*}\delta(t) = \infty$. Therefore

$$\left[0, C^{-1}\log(1+\delta(0)^{-1})\right) \subset \mathrm{dom}\psi_\xi\,.$$

In a similar way we can prove that

$$\left(C^{-1}\log(1+\delta(0)^{-1}), 0\right] \subset \mathrm{dom}\psi_\xi$$

Since $\delta(0) = \|\psi_\xi(0) - \psi_x(0)\| = \|\xi - x\|$ and $x \in E$ was arbitrary, by making x range over D we get the first statement; the second one has been proved in (7.13).

#

7.36. DEFINITION. *For* $\tau \in \mathbb{R}$ *we set*

$$D(\tau, A) =: \{\xi \in E;\ \tau \in \mathrm{dom}\psi_\xi\}$$

where ψ_ξ *denotes the maximal solution of (7.11).*

Thus, we have $D \subset D(\tau, A)$ for all $\tau \in \mathbb{R}$

7.37. LEMMA. *The following relations hold*

(a) $\{\xi \in E;\ \mathrm{dist}(\xi, D) < (e^{C|\tau|} - 1)^{-1}\} \subset D(\tau, A)$ *for all* $\tau \in \mathbb{R}$.

(b) *If* $A \in \mathrm{aut}D$, *then* $A_{|\partial D} \in \mathrm{aut}(\partial D)$ *and* $(\exp tA)\partial D = \partial D$ *for all* $t \in \mathbb{R}$.

Proof: Let $\tau \in \mathbb{R}$ be given and assume that $\xi \in E$ satisfies $\mathrm{dist}(\xi, D) < (e^{C|\tau|} - 1)^{-1}$. We may suppose that $\xi \notin D$ as otherwise there is nothing to prove. From

$$e^{C|\tau|} < 1 + \mathrm{dist}(\xi, D)^{-1}$$

we obtain

$$|\tau| < C^{-1} \log[1 + \mathrm{dist}(\xi, D)^{-1}]$$

so that, by lemma 7.35 we obtain $\tau \in \mathrm{dom}\psi_\xi$ and $\xi \in D(\tau, A)$.

The above reasoning also shows that, for all $\xi \in \partial D$, we have $\xi \in D(\tau)$ whatever is $\tau \in \mathbb{R}$. Moreover, due to the uniqueness of the solution of the initial value problem (7.11),

$$(\exp tA)\partial D \subset E \setminus D$$

for all $t \in \mathbb{R}$. We shall now show that

$$(\exp tA)\partial D \subset \bar{D}$$

Therefore, for $t \in \mathbb{R}$ we have

(7.14) $$(\exp tA)\,\partial D \subset \partial D$$

Indeed, let us fix any $\xi \in \partial D$ and $\tau \in \mathbb{R}$. Then, for every $\varepsilon>0$, there exists some $x_\varepsilon \in D$ such that

$$\| \xi - x_\varepsilon \| < [(1+\varepsilon^{-1})e^{C|\tau|-1}-1]^{-1}$$

whence

$$|\tau| < C^{-1}\log[1+\| \xi - x_\varepsilon \|^{-1}]$$

Applying lemma 7.35 to x_ε and τ we get

$$\delta(\tau) = \| \psi_\xi(\tau) - \psi_{x_\varepsilon}(\varepsilon) \| \leq [(1+\| \xi - x_\varepsilon \|^{-1})e^{-C|\tau|}-1]^{-1}$$

For $\varepsilon \to 0$ we have $\| \xi - x_\varepsilon \| \to 0$; therefore $\| \psi_\xi(\tau) - \psi_{x_\varepsilon}(\tau)\| \to 0$. Since $\psi_{x_\varepsilon}(\tau) \in D$, we have $\psi_\xi(\tau) \in \bar{D}$. As $\xi \in \partial D$ and $\tau \in \mathbb{R}$ were arbitrary, we get $(\exp tA)\,\partial D \subset \bar{D}$ for $t \in \mathbb{R}$.

Finally, applying $\exp(-tA)$ to (7.14) we obtain $\partial D \subset [\exp(-tA)]\,\partial D \subset \partial D$ so that $(\exp tA)\,\partial D = \partial D$ for all $t \in \mathbb{R}$.

#

As usually, we write D_ε for the parallel set $D_\varepsilon =: D + \varepsilon B(E)$.

7.38. LEMMA. *Let* $A \in \operatorname{aut} D$ *be given. Then, for every* $\varepsilon>0$, *there exists* $\eta>0$ *such that* $f =: \exp A$ *admits an injective holomorphic extension* $F\colon D_\eta \to E$ *with* $F(D_\eta) \subset D_\varepsilon$. *In particular,* F *is bounded on* D_η.

Proof: Let $A \in \operatorname{aut} D$ be given and denote by ρ and C the numbers associated with A by definition 7.34. We may assume that $C>0$ as otherwise the statement is obviously true. By lemma 7.37, once $\tau \in \mathbb{R}$ has been fixed, we have

$$\{\xi \in E;\ \operatorname{dist}(\xi, D) < (e^{C|\tau|}-1)^{-1}\} \subset \{\xi \in E;\ \tau \in \operatorname{dom}\psi_\xi\}$$

thus, by setting $\tau = 1$ and

$$\eta = \min\{(e^{C}-1)^{-1},\ \ [(1+\varepsilon^{-1})e^{C}-1]^{-1}\} > 0$$

we have

$$D_\eta \subset \text{dom expA}$$

so that it makes sense to define F: $D_\eta \to E$ by means of $F(\xi) =: (\exp A)\xi$; that is, $F(\xi)$ is the value at the instant t= 1 of the solution of the initial value problem

$$\frac{dy}{dt} = A[y(t)], \quad y(0) = \xi$$

From the theory of ordinary differential equations we know that F: $D_\eta \to E$ is an injective holomorphic mapping. Obviously, we have $F_{|D} = f$. By applying lemma 7.35 to any $x \in D$, any $\xi = x+u$ with $u \in E$, $\|u\| < \eta$, and t= 1 we obtain

$$\|F(x+u)-F(x)\| < [(1+\|u\|^{-1})e^{-C}-1]^{-1} < [(1+\eta-1)e^{-C}-1] < \varepsilon$$

Whence $F(D_\eta) \subset D_\varepsilon$. The extension F of f= expA is clearly unique. #

7.39. THEOREM. *Let* $f \in \text{Aut}D$ *be given. Then, for every* $\varepsilon > 0$, *there exist* $\eta > 0$ *such that* f *admits and injective holomorphic extension* F: $D_\eta \to E$ *with* $F(D_\eta) \subset D_\varepsilon$. *In particular,* F *is bounded on* D_η.

Proof: Let $f \in \text{Aut}D$ and $\varepsilon > 0$ be given. By theorem 7.28 we have $f = g \circ h$ for some $g \in \text{Aut}^0 D$ and $h \in \text{Aut}_0 D$. Then, by theorem 7.25 and lemma 7.12 it follows that

$$f = G_{|D} \circ (\exp A_1) \circ \ldots \circ (\exp A_n)$$

for some surjective linear isometry G: $E \to E$ and $A_1, \ldots A_n \in \text{aut}D$. Obviously $g = G_{|D}$ admits an injective holomorphic extension to the whole E. Moreover, as G is uniformly continuous on E, we can easily find a number $\eta_0 > 0$ such that

$$G(D_{\eta_0}) \subset D_\varepsilon$$

Now, by lemma 7.38 we can find a number $\eta_1 > 0$ such that $h_1 =: \exp A_1$ admits an injective holomorphic extension H_1 to D_{η_1}

with

$$H_1(D_{\eta_1}) \subset D_{\eta_0}$$

By reiterating the argument and setting $\eta =: \min(\eta_0, \eta_1, \ldots, \eta_n)$ we get the result.

#

<u>7.40. EXERCISES</u>. (a) Show that, for any $f \in \mathrm{Aut}D$, there is a parallel negighbourhood D_η of D in which f is uniformly continuous.

(b) Show that any $f \in \mathrm{Aut}D$ maps any subset $S \subset\subset D$ into a subset $f(S) \subset\subset D$.

CHAPTER 8

AUTOMORPHISMS OF THE UNIT BALL OF SOME CLASSICAL BANACH SPACES

§1.- Some geometrical considerations.

Let A be a convex non void subset of E. We recall that a function $f\colon A\to \mathbb{R}$ is "convex" if we have

$$f[\lambda+(1-\lambda)y]\leqslant\lambda f(x)+(1-\lambda)f(y)$$

for all $\lambda\in[0,1]$ and all $x,y\in A$.

We denote by Conv(A) the set of the functions $f\colon A\to \mathbb{R}$ that are convex in A.

8.1. DEFINITION. *We say that* $f\in\mathrm{Conv}(A)$ *is "subdifferentiable" at a point* $a\in A$ *if there exists a continuous real linear functional* $\Lambda\colon E\to\mathbb{R}$ *such that we have*

$$\Lambda(x)-\Lambda(a)\leqslant f(x)-f(a)$$

for all $x\in A$.

The set $\nabla f(a)$ of the functionals Λ in those conditions is called the *subgradient* of f at the point a.

8.2. PROPOSITION. *Any lower semicontinuous function is subdifferentiable. In particular, the norm* $\|\cdot\|$ *is subdifferentiable at any point* $a\in E$, $a\neq 0$, *and we have*

$$\nabla\|\cdot\|(a)=\{\Lambda\in E^*_{\mathbb{R}};\ \|\Lambda\|=1,\ \ \Lambda(a)=\|a\|\}$$

Here $E^*_{\mathbb{R}}$ denotes the set of continuous real linear functionals $\Lambda: E \to \mathbb{R}$.

Proof: See |28| theorem F page 30.

Since any continuous complex linear functional $\phi \in E^*_{\mathbb{C}}$ is uniquely determined by its real part $\Lambda \in E^*_{\mathbb{R}}$ by the formula

$$\phi(x) = \Lambda(x) - i\Lambda(ix), \quad x \in E,$$

and $\|\phi\| = \|\Lambda\|$, we also have

$$\nabla\|\cdot\|(a) = \{\mathrm{Re}\phi;\ \phi \in E^*_{\mathbb{C}},\ \|\phi\| = 1,\ \phi(a) = \|a\|\}$$

Let $D =: B(E)$ be the open unit ball of E. We recall from chapter 7 that any $X \in \mathrm{aut} D$ can be uniquely represented in the form

$$X(x) = c + L(x) - Q_c(x,x), \quad x \in E$$

where $c \in E_0$, $L \in \mathcal{L}(E)$ and $Q_c \in \mathcal{L}_s(E \times E|E)$ is a continuous symmetric bilinear map.

<u>8.3. PROPOSITION</u>. *Let* $a \in \partial D$ *and* $X \in \mathrm{aut} D$ *be given. Then, for all* $\phi \in E^*_{\mathbb{C}}$ *such that* $\mathrm{Re}\phi \in \nabla\|\cdot\|(a)$ *we have*

$$\mathrm{Re}\phi[L(a)] = 0 \qquad \phi[Q_c(a,a)] = \overline{\phi(c)}$$

Proof: Let us fix any $\theta \in \mathbb{R}$. We have $e^{i\theta}a \in \partial D$. Consider the initial value problem

$$\frac{d}{dt}\, y(t) = X[y(t)], \qquad y(0) = e^{i\theta}a$$

and denote by $\varphi(t) =: (\exp tX)e^{i\theta}a$ its maximal solution. By lemma 7.37 we know that $X|_{\partial D}$ is complete in ∂D so that

$$\mathrm{dom}\,\varphi = \mathbb{R}, \qquad \varphi(t)\partial D = \partial D$$

for all $t \in \mathbb{R}$. Let us define

$$\psi(t) =: e^{-i\theta}.\exp(tX)e^{i\theta}a$$

for $t\in\mathbb{R}$. Then

$$\psi(0) = a, \qquad \|\psi(t)\| = 1$$

for $t\in\mathbb{R}$. Thus, if $Re\,\phi\in\nabla\|\cdot\|(a)$,

$$Re[\psi(t)] - Re[\psi(0)]| \leq \|\psi(t)\| - \|\psi(0)\| = 0$$

for $t\in\mathbb{R}$. Dividing by t and letting $t\downarrow 0$ we obtain

$$Re\,\phi[\psi'(0)] \leq 0$$

In a similar manner, by setting

$$\psi(t) =: e^{-i\theta}.\exp(-tX)e^{i\theta}a$$

we obtain

$$Re\,\phi[\psi'(0)] \geq 0$$

so that we have $Re[\psi'(0)] = 0$. Substitution of

$$\psi'(0) = e^{-i\theta}\,\varphi'(0) = e^{-i\theta}X[\varphi(0)] = e^{-i\theta}X(e^{i\theta}a) =$$
$$= e^{-i\theta}c + L(a) - e^{-i\theta}Q_c(a,a)$$

gives

$$Re[e^{-i\theta}c + L(a) - e^{i\theta}Q_c(a,a)] = 0$$

But this is valid for all $\theta\in\mathbb{R}$, whence the result follows immediately. #

<u>8.4. COROLLARY</u>. *Let* $X\in\text{aut}D$, $a\in E$ *and* $\phi\in E^*_{\mathbb{C}}$ *be such that* $\phi(a) = \|\phi\|\cdot\|a\|$. *Then we have*

$$Re\,\phi[L(a)] = 0 \quad , \qquad \phi[Q_c(a,a)] = \|a\|^2\overline{\phi(c)}$$

Proof: We may assume $a\neq 0$ and $\phi\neq 0$ as otherwise the result is obvious. From $\phi(a) = \|\phi\|\;\|a\|$ we obtain

$$Re\,\frac{\phi}{\|\phi\|} \in \nabla\|\;\|\left(\frac{a}{\|a\|}\right)$$

Then proposition 8.2 applied to $\frac{\phi}{\|\phi\|}$ and $\frac{a}{\|a\|}$ gives the result.

#

§2.- Automorphisms of the unit ball of $L^p(\Omega,\mu)$, $2\neq p\neq\infty$.

Throughout this section, (Ω,μ) represents a real measure space with $0<\mu(\Omega)<\infty$. For each p with $1\leqslant p<\infty$, L^p stands for the Banach space of (classes of) measurables functions $f:\ \Omega\rightarrow\mathbb{C}$ such that $\|f\|_p =: (\int_\Omega |f|^p\, d\mu)^{1/p}<\infty$. We shall make no distinction between f and its corresponding class. With q we represent the conjugate of p, i.e. $\frac{1}{p}+\frac{1}{q}=1$. The usual conventions are applicable for the case $p=\infty$. For each measurable set S, χ_S denotes the characteristic function of S.

8.5 LEMMA. *Let* p *be given with* $1\leqslant p<\infty$ *and assume that* $\dim^p L>1$. *Then, there are two measurable sets,* A, B *such that*

$$A\cap B=\phi\ ,\quad A\cup B=\Omega\ ,\quad 0<\mu(A)<\mu(\Omega)\ ,\quad 0<\mu(B)<\mu(\Omega)\,.$$

Proof: We shall show that there exists a measurable set A satisfying $0<\mu(A)<\mu(\Omega)$. Then the pair A and $B=:\ \Omega\setminus A$ satisfies the requirements of the lemma.

We proceed by contradiction. Assume that for every measurable set A we had

$$\text{(8.1)}\qquad \text{either}\quad \mu(A)=0\quad \text{or}\quad \mu(A)=\mu(\Omega)$$

Since $\mu(\Omega)<\infty$, we have $\chi_\Omega\in L^p$. As $\dim L^p>1$, there exists some $f\in L^p$ which does not belong to the subspace generated by χ_Ω; in particular

$$\text{(8.2)}\qquad \mu(\{x\in\Omega;\ f(x)\neq 0\})>0$$

Obviously, we have

$$\text{(8.3)}\qquad \{x\in\Omega;\ f(x)\neq 0\}=(\bigcup_{n\geqslant 0} f^{-1}(n,n+1])\cup(\bigcup_{n<0} f^{-1}(n-1,n])$$

Thus, by (8.1) and (8.2), exactly one of the subsets in the right-hand side of (8.3) has positive measure. We may assume the corresponding subinterval to be $I_0 = (0,1]$ so that

$$A_0 =: f^{-1}(0,1] \quad , \quad 0<\mu(A_0) = \mu(\Omega) \tag{8.4}$$

Now we write $A_0 = f^{-1}(0,1/2] \cup f^{-1}(1/2,1]$. Again, exactly one of these two subsets has positive measure. We write I_1 for the corresponding subinterval; thus

$$A_1 = f^{-1}(I_1), \qquad 0<\mu(A_1) = \mu(\Omega)$$

By induction we get a sequence of intervals $(I_k)_{k\in\mathbb{N}}$ and a sequence of measurable sets $(A_k)_{k\in\mathbb{N}}$ with

$$A_k =: f^{-1}(I_k) \quad , \quad 0<\mu(A_k) = \mu(\Omega)$$

for all $k\in\mathbb{N}$. As the sequences of the extreme points of the I_k are monotonous and bounded, they define a number $\xi\in\mathbb{R}$ and it is easy to check that

$$\bigcap_{k\in\mathbb{N}} A = \{x\in\Omega;\ f(x) = \xi\}$$

Moreover, as $\mu(\Omega)<\infty$,

$$\mu(\bigcap_{k\in\mathbb{N}} A_k) = \lim_{k\to\infty} \mu(A_k) = \mu(\Omega)$$

Therefore,

$$\mu(\{x\in\Omega;\ f(x)\neq\xi\}) = 0$$

and we have $f = \xi\chi_\Omega$ almost everywhere; thus f belongs to the subspace generated by χ_Ω , which is a contradiction.

#

8.6. DEFINITION. *Let* p *with* $1\leq p<\infty$ *be given. For* $f\in L^p$, *we define* f^* *in the following manner*

(a) *If* $p=1$, *then* $f^*(x)=\dfrac{\bar{f}(x)}{|f(x)|}\,\|f\|_1$ *when* $f(x)\neq 0$ *and* $f^*(x)=0$ *when* $f(x)=0$.

(b) *If* $p\neq 1$, *then* $f^*=\bar{f}|f|^{p-2}$.

<u>8.7. LEMMA</u>. *For* $f\in L^p$, *the function* f^* *satisfies*

(a) $f^*\in L^q$

(b) $\|f\|_p^p=\|f^*\|_q^q$

(c) $\langle f,f^*\rangle=\|f\|_p\,\|f^*\|_q$.

Proof: First we consider the case $p=1$. Obviously f^* is essentially bounded and $\|f^*\|_\infty=\|f\|_1$. Moreover, writing $S=:\{x\in\Omega;\ f(x)=0\}$ we have

$$\langle f,f^*\rangle=:\int_\Omega f.f^*d\mu=\int_S f.f^*d\mu+\int_{\Omega\setminus S} f.f^*d\mu=\int_{\Omega\setminus S} f\,\frac{\bar{f}}{|f|}\,\|f\|_1 d\mu=$$

$$=\|f\|_1\int_{\Omega\setminus S}|f|d\mu=\|f\|_1\|f\|_1=\|f\|_1\|f^*\|_\infty$$

Now we consider the case $1<p<\infty$. Then

$$|f^*|^q=|\bar{f}|^q(|f|^{p-2})^q=|\bar{f}|^q|f|^{pq-2q}=|f|^{pq-q}=|f|^p$$

Thus

$$\|f\|_p^p=\int_\Omega|f|^p\,d\mu=\int_\Omega|f^*|^q\,d\mu=\|f^*\|_q^q$$

Finally, since $p\neq 1$,

$$\langle f,f^*\rangle=\int_\Omega f.f^*d\mu=\int_\Omega f.\bar{f}|f|^{p-2}d\mu=\int_\Omega|f|^p d\mu=\|f\|_p^p=\|f\|_p\|f\|_p^{p-1}=$$

$$=\|f\|_p(\|f\|_p^p)^{1/q}=\|f\|_p\|f^*\|_q.$$

#

8.8 LEMMA. *Let* p, $1\leq p<\infty$, $\rho\geq 0$ and $\theta\in\mathbb{R}$ *be given. Then, for each pair of measurable sets* A,B *with* $A\cap B=\phi$, $A\cup B=\Omega$, *the function* $f=:\chi_A+\rho\, e^{i\theta}\chi_B$ belongs to L^p *and its corresponding* f^* *is given by*

$$f^*=\chi_A+\rho^{p-1}e^{-i\theta}\chi_B \quad \textit{if } p\neq 1$$

$$f^*=[\mu(A)+\rho\mu(B)](\chi_A+e^{-i\theta}\chi_B) \quad \textit{if } p=1$$

Proof: It is an easy consequence of the definitions and the facts $\chi_A\cdot\chi_B=0$, $\chi_A+\chi_B=1$.

#

8.9. THEOREM. *Let* p *be given with* $1\leq p<\infty$, $p\neq 2$ *and assume that* $\dim L^p>1$. *Then every vector field that is complete in the unit ball of* L^p *is linear.*

Proof: Let $X\in\mathrm{aut}B(L^p)$ be given. We recall that there exists some $c\in L^p$ such that X admits the representation

$$X(x)=c+L(x)-Q_c(x,x), \qquad x\in L^p$$

First, we shall prove the following auxiliary statement.

8.10. LEMMA. *Under the assumptions of theorem 8.9, for every measurable set* A *with* $0<\mu(A)<\mu(\Omega)$ *we have*

$$\int_A c\,d\mu=0$$

Proof: Once A has been given, we put $B=:\Omega\setminus A$. In order to simplify the notations, we set

$$\mu_1=:\mu(A) \qquad \mu_2=:\mu(B) \qquad \chi_1=:\chi_A \qquad \chi_2=:\chi_B$$

$$\gamma_j=:\langle\alpha,\chi_j\rangle \qquad \beta^{\ell}_{j,k}=:\langle Q_\alpha(\chi_j,\chi_k);\chi_\ell\rangle$$

for $j,k,\ell=1,2$. Now, by lemma 8.8, for $\rho\geq 0$ and $\theta\in\mathbb{R}$ we have $f=:\chi_1+\rho\, e^{i\theta}\chi_2\in L^p$. By lemma 8.7 we have $\langle f,f^*\rangle=\|f\|_p\|f^*\|_q$;

thus, applying corollary 8.4 to the pair (f,f*), we obtain

$$\|f\|_p^2 \langle c, f^* \rangle = \langle Q_c(f,f);\ f^* \rangle$$

Therefore, from the expression of f* given by lemma 8.8 we get in the case $p \neq 1$

$$(\mu_1+\rho^2\mu_2)^{2/p} \overline{\langle c, \chi_1+\rho^{p-1}\ e^{-i\theta}\ \chi_2 \rangle} =$$

$$= \langle Q_c(\chi_1+\rho\ e^{i\theta}\ \chi_2;\ \chi_1+\rho\ e^{i\theta}\ \chi_2),\ \chi_1+\rho^{p-1}\ e^{-i\theta}\ \chi_2 \rangle$$

After computing the integrals appearing in this formula and rearranging the terms in accordance with the powers of $e^{i\theta}$, we get

$$\beta^1_{22}\ \rho^2\ e^{2i\theta} + [\beta_{22}\ \rho^{p+1} + 2\beta^1_{12}\ \rho + (\mu_1+\rho^2\mu_2)^{2/p}\ \bar{\gamma}_2\ \rho^{p-1}] +$$

$$+ [\beta^1_{11} + 2\beta^2_{12}\ \rho^p + (\mu_1+\rho^p\ \mu_2)^{2/p}\ \bar{\gamma}_1] + \beta^2_{11}\ \rho^{p-1}\ e^{i\theta} = 0$$

which is valid for all $\rho \geq 0$ and $\theta \in \mathbb{R}$. Thus, all coefficients must be null; in particular

$$\text{(8.5)} \qquad \lim_{\rho\to\infty} \frac{1}{\rho^2} [\beta^1_{11} + 2\beta^2_{12}\ \rho^p + (\mu_1+\rho^p\ \mu_2)^{2/p}\ \bar{\gamma}_1] = 0$$

Besides, as β^1_{11} is independent of ρ,

$$\text{(8.6)} \qquad \lim_{\rho\to\infty} \frac{1}{\rho^2}\ \beta^1_{11} = 0$$

Also,

$$\text{(8.7)} \qquad \lim_{\rho\to\infty} \frac{1}{\rho^2} (\mu_1+\rho^p\ \mu_2)^{2/p}\ \bar{\gamma}_1 = \bar{\gamma}_1\ \mu_2^{2/p}$$

From (8.5), (8.6) and (8.7) we derive

$$\lim_{\rho\to\infty} 2\beta_{12}^2 \rho^{p-2} = \bar{\gamma}_1 \mu_2^{2/p}$$

As we have assumed $p\neq 2$ and β_{12}^2 does not depend on ρ, the last equality entails $\beta_{12}^2 = 0$ and $\bar{\gamma}_1 = 0$. Therefore

$$0 = \gamma_1 = \langle c, \chi_1 \rangle = \int_\Omega c\, \chi_1 d\mu = \int_A c d\mu$$

as we wanted to show. The proof in the case $p = 1$ is similar.

Now we we proceed to prove the theorem. Obviously, it suffices to show that, for any measurable set S,

$$\int_S c d\mu = 0.$$

This is clear when $\mu(S) = 0$. If $0<\mu(S)<\mu(\Omega)$, the result is also true by lemma 8.10. Thus we have to consider the case $\mu(S) = \mu(\Omega)$. Since we have assumed that $\dim L^p > 1$, by lemma 8.5 there are measurable sets A,B with $A\cap B = \phi$, $A\cup B = \Omega$, $0<\mu(A)<\mu(\Omega)$, $0<\mu(B)<\mu(\Omega)$. From lemma 8.10 applied to A and B we get

$$\int_S c d\mu = \int_\Omega c d\mu = \int_A c d\mu + \int_B c d\mu = 0 \qquad \#$$

8.11. COROLLARY. *For* $1\leq p<\infty$, $p\neq 2$ *and* $\dim L^p > 1$ we have $(\mathrm{Aut} D)0 = \{0\}$ *and*

$$\mathrm{Aut} D = \{U_{|D} \;;\; U \textit{ is a surjective linear isometry of } L^p\}.$$

Proof: We recall from definition 7.8 and proposition 7.23 that if $E_0 = \{X(0)\,;\, X\in \mathrm{aut} D\}$, then we have $(\mathrm{Aut} D)0 = E_0 \cap D$. From theorem 8.9 we obtain $E_0 = \{0\}$, thus $(\mathrm{Aut} D)0 = \{0\}$. Then, for $F\in \mathrm{Aut} D$ we have $F(0) = 0$; therefore, by corollary 7.25, $F = U_{|D}$ for some surjective linear isometry U of L^p. The converse is obvious.

#

§3.- Automorphisms of the unit ball of some algebras of continuous functions.

Throughout this section Ω denotes a compact Hausdorff space and $C(\Omega)$ represents the Banach space of continuous functions $f: \Omega \to \mathbb{C}$ with the norm $\|f\| =: \sup_{x\in\Omega} |f(x)|$.

8.12. THEOREM. *Let* A *be closed complex subalgebra of* $C(\Omega)$ *which contains the unit* e *of* $C(\Omega)$. *Suppose that* $T: A \to A$ *is a surjective linear isometry of* A. *Then* T *is of the form*

$$Tf = \alpha . \phi f \qquad f\in A$$

where α *is an element of* A *such that* $|\alpha(x)| = 1$ *for all* $x\in\Omega$, $\frac{1}{\alpha} \in A$ *and* ϕ *is a surjective algebra automorphism of* A. *Moreover, if* $Te = e$, *then* T *is multiplicative*

Proof: We consider A as a Banach space and write A^* for its conjugate space. Thus, the unit ball $B(A^*)$ is a convex set that is compact for the weak-* topology of A^*. By the Krein-Milman theorem $\bar{B}(A^*)$ is the weak-* closed convex hull of the set extr $\bar{B}(A^*)$ of the extreme points of $B(A^*)$. In particular, extr $\bar{B}(A^*) \neq \phi$.

First we show that every extreme point L of $B(A^*)$ is of the form $L = \lambda\delta_x$ where $\lambda\in\mathbb{C}$, $|\lambda| = 1$, and $\delta_x: A \to \mathbb{C}$ is the evaluation $f \to f(x)$ at some point $x\in\Omega$. Indeed, let L be an extreme point of $B(A^*)$; then $\|L\| = 1$ and, by the Hahn-Banach theorem, we can extend L to an element $F\in C(\Omega)^*$ such that $\|F\| = 1$. Let S denote the set of such extensions

$$S =: \{F\in C(\Omega)^*;\ F_{|A} = L,\ \ \|F\| = \|L\| = 1\}$$

It is clear that S is a convex and weak-* compact subset of $C(\Omega)^*$; thus, using again the Krein-Milman theorem, we can choose an extreme point F of S. Then F is actually an extreme point of the unit ball of $C(\Omega)^*$. Indeed, if we have

$$F = \frac{1}{2}(F_1 + F_2)$$

with $\|F_k\| \leq 1$, k= 1,2, then

$$\|F_1\| = \|F_2\| = 1$$

and, if we write L_k for the restriction of F_k to A, then we have $\|L_k\| \leq 1$ and

$$L = \frac{1}{2}(L_1+L_2)$$

Since L is an extreme point of B(A*) we get $L_1 = L_2 = L$. Therefore, each F_k is a norm preserving extension of L, i.e., $F_k \in S$, and because F is an extreme point of S, we have $F_1 = F_2 = F$. Now, as $C(\Omega)^*$ is the space of complex Baire measures on Ω, the extreme points of its unit ball are known (cf. |13|) to be the measures of the form $\lambda\delta_x$ where $\lambda \in \mathbb{C}$, $|\lambda| = 1$ and δ_x is the evaluation on $C(\Omega)$ at some point $x \in \Omega$. It follows that our extreme functional F has the form

$$F(f) = \lambda f(x) \qquad f \in C(\Omega)$$

and, restricting to A, we obtain

$$L(f) = \lambda f(x) \qquad f \in A$$

so that $L = \lambda\delta_x$.

As a well known consequence of the Hanh-Banach theorem, we have

$$\|f\| = \max\{|Lf|;\ L \in \bar{B}(A^*)\} \qquad f \in A$$

Given $f \in A$, the functional $L \to |Lf|$ is convex and weak-* continuous on A*. So, it attains its maximum on the weak-* compact convex set $\bar{B}(A^*)$ at some extremal point (cf. |28|)

$$\|f\| = \max\{|Lf|;\ L \in \text{extr } \bar{B}(A^*)\} \tag{8.8}$$

Now we consider the given Banach space isometry T of A. The transposed mapping T* of T, defined by

$$(T^*L)f = L(Tf) \qquad L\in A^*, \quad f\in A$$

is a surjective isometry of A^*. Thus T^* must carry the set $\operatorname{extr}B(A^*)$ onto itself. Let us define

$$\Omega_0 =: \{x\in\Omega; \quad \delta_x \in \operatorname{extr}B(A^*)\}$$

Since $\delta_x \in \operatorname{extr}B(A^*)$ if and only if $\lambda\delta_x \in \operatorname{extr}B(A^*)$ for all $\lambda\in\mathbb{C}$, $|\lambda| = 1$, we have

$$\operatorname{extr}B(A^*) = \{\lambda\delta_x; \ x\in\Omega_0, \ |\lambda| = 1\} \tag{8.9}$$

If $x\in\Omega_0$, then $T^*\delta_x$ must also be an extreme point of $B(A^*)$. This associates with x a complex number $\alpha(x)$ of modulus 1 and a point $\tau(x)\in\Omega_0$ such that

$$T^*\delta_x = \alpha(x)\delta_{\tau(x)}$$

or, equivalently,

$$(Tf)(x) = \alpha(x).f[\tau(x)] \tag{8.10}$$

for all $f\in A$ and $x\in\Omega_0$. Taking $f = e\in A$ in the above, we get $\alpha(x) = (Te)(x)$ for all $x\in\Omega_0$ so that α is (the restriction to Ω_0 of) a function $Te\in A$.

Now we begin to use the ring structure of A. Let $f, g\in A$ and $x\in\Omega_0$ be given. Then we have

$$[T(fg)](x) = \alpha(x)(fg)[\tau(x)] = \alpha(x)f[\tau(x)]g[\tau(x)]$$

and multiplying by $\alpha(x)$ we obtain

$$\alpha(x)[T(fg)](x) = (Tf)(x)(Tg)(x) \tag{8.11}$$

From (8.8) and (8.9) we derive that, for any $h\in A$,

$$\|h\| = \sup\{|Lh|; \ L\in\operatorname{extr}B(A^*)\} = \sup_{x\in\Omega_0} |h(x)|$$

From (8.11) we see that, for $f, g \in A$, the function

$$h =: \alpha T(fg) - (Tf)(Tg)$$

vanishes on Ω_0; consequently

$$\alpha T(fg) = (Tf)(Tg)$$

for all $f, g \in A$. If we take $f = g = T^{-1}(e)$, we see that $\frac{1}{\alpha} \in A$. Since both α and $\frac{1}{\alpha}$ have modulus 1 on Ω_0 and each attains its maximum on Ω, we see $|\alpha(x)| = 1$ for all $x \in \Omega$. Then if we define $\phi: A \to A$ by

$$\phi f = \alpha^{-1} Tf \qquad f \in A$$

it is clear that ϕ is a surjective algebra automorphism of A.

#

Next we characterize the group of automorphisms of the unit ball $D =: B(A)$ of the algebra A. Since D is bounded circular domain, there is a Banach subspace A_0 associated with A by definition 7.5. We recall from chapter 7 that any vector field $X \in \operatorname{aut} D$ can be uniquely represented in the form

$$X(h) = f + L(h) - Q_f(h,h) \qquad h \in A$$

for some $L \in \operatorname{aut}^0 D$, $f - Q_f \in \operatorname{aut}_0 D$ and $f \in A_0$.

<u>8.13. THEOREM.</u> *Let Ω be a compact Hausdorff space and denote by* A *a closed complex subalgebra of* $C(\Omega)$ *such that* $e \in A$. *Then*

(a) A_0 *is the complex subalgebra of* A *given by* $A_0 =: \{a \in A;\ \bar{a} \in A\}$.

(b) *Every automorphism* $F \in \mathrm{Aut}D$ *admits a unique representation* $F = L \circ M$ *where* $L \in \mathrm{Aut}^0 D$ *is a surjective linear isometry of* A *and* $M \in \mathrm{Aut}_0 D$ *is the Möbius transformation*

$$Mf =: \frac{f+a}{e+\bar{a}f} \qquad f \in D$$

for some $a \in A_0$, $\|a\| < 1$.

Proof: Let us write $R(A) =: \{a \in A;\ a = \bar{a}\}$ and notice that $R(A)$ is a closed real subalgebra of A with $e \in R(A)$. Moreover, its complexified $R(A)^{\mathbb{C}}$ satisfies

$$R(A)^{\mathbb{C}} = \{a \in A; \quad \bar{a} \in A\}$$

It is easy to see that, for $a \in R(A)^{\mathbb{C}}$ with $\|a\| < 1$, the Möbius transformation

$$Mf = \frac{f+a}{1+\bar{a}f}, \qquad f \in D$$

is an element of AutD. Thus

$$R(A)^{\mathbb{C}} \cap D \subset (\mathrm{Aut}D)0$$

and, by proposition 7.23

$$(8.12) \qquad R(A)^{\mathbb{C}} \subset A_0$$

Now let $f \in A_0$ be given. Then, the vector field $X_f(h) =: f - Q_f(h,h)$, $h \in A$, satisfies $X \in \mathrm{aut}D$. If $\delta_x : A \to \mathbb{C}$ denotes the evaluation on A at any point $x \in \Omega$, we have $\delta_x \in A^*$ and

$$1 = \delta_x(e) = \|\delta_x\| \ \|e\|$$

As $e\in\partial D$, from proposition 8.3 we obtain

$$(8.13) \qquad \delta_x[Q_f(e,e)] = \overline{\delta_x f} = \delta_x \bar{f} \qquad x\in\Omega$$

so that $Q_f(e,e) = \bar{f}$ and therefore $\bar{f}\in A$. Thus

$$A_0 \subset R(A)^{\mathbb{C}}$$

From (8.12) and (8.13) we get

$$A_0 = R(A)^{\mathbb{C}} = \{a\in A;\quad \bar{a}\in A\}.$$

The remainder part of the proof follows easily from theorem 7.25.

#

A number of interesting examples are included in theorem 8.13

8.14. EXAMPLE. *Let* A *be the whole algebra* $C(\Omega)$. *Then the group* AutD *can be represented as the set of the transformations* $F: f \to Ff$ *given by*

$$(8.14) \qquad Ff = \alpha . \frac{f\circ\tau + a\circ\tau}{e + (\bar{a}\circ\tau).(f\circ\tau)}, \qquad f\in D$$

where $\tau: \Omega \to \Omega$ is a homeomorphism of Ω, $\alpha: \Omega \to \mathbb{C}$ is a continuous function with $|\alpha(x)| = 1$ for all $x\in\Omega$ and $a\in C(\Omega)$ with $\|a\| < 1$.

Proof: By theorem 8.13 we have $A_0 = \{f\in C(\Omega);\ \bar{f}\in C(\Omega)\} = C(\Omega)$; thus, by proposition 7.23, the orbit of the origin is

$$(\mathrm{Aut}D)0 = C(\Omega) \cap D = D$$

Therefore, any $F\in\mathrm{Aut}D$ can be uniquely represented in the form $F = L\circ M$ where M is a Möbius transformation and L is a surjective linear isometry of $C(\Omega)$. But it is known (cf. |13|) that any L in those conditions admits a unique representation

$$(Lf)x = \alpha(x).f[\tau(x)] \qquad x\in\Omega, \quad f\in C(\Omega)$$

for some homeomorphism τ of Ω and some $\alpha\in C(\Omega)$ with $|\alpha(x)| = 1$

for all $x\in\Omega$, whence we obtain (8.14). The converse is obvious.

#

<u>8.15. EXAMPLE</u>. *Let* $\Omega =: \bar{\Delta}$ *be the closed unit disk of* $\mathbb{C}$ *and denote by* A *the algebra of all functions* $f: \bar{\Delta} \to \mathbb{C}$ *that are continuous on* $\bar{\Delta}$ *and holomorphic on* Δ. *Then* AutD *is the set of the transformations*

$$(8.15) \qquad Ff = \alpha . \frac{f\circ\tau + a}{e+\bar{a}.(f\circ\tau)} \qquad f\in D$$

where $\tau\in \mathrm{Aut}\Delta$ *is a conformal map of the unit disk and* α, a *are complex numbers with* $|\alpha| = 1$, $|a|<1$.

Proof: By theorem 8.13 we have $A_0 = \{f\in A;\ \bar{f}\in A\}$. For a function $f\in H(\Delta)$, its conjugate $\bar{f}$ is holomorphic on Δ if and only if f is constant; thus, in our case $A_0 = \mathbb{C}$ and the orbit of the origin is

$$(\mathrm{Aut}D)0 = \mathbb{C} \cap D = \Delta$$

Then, any $F\in \mathrm{Aut}D$ can be uniquely represented in the form $F = L\circ M$ where M is a transformation

$$Mf = \frac{f+a}{e+\bar{a}f}$$

with $a\in\Delta$ and L is a surjective linear isometry of A. By theorem 8.12, such an L can be uniquely represented in the form

$$Lf = \alpha.\phi f \qquad f\in A$$

where $\alpha\in A$ is a function satisfying $|\alpha(x)| = 1$ for all $x\in\Delta$ and ϕ is an algebra automorphism of A. Since in our case α is holomorphic on Δ and has constant modulus, α is constant. Thus, the result follows from the proposition below.

<u>8.16. PROPOSITION</u>. Let ϕ *be a surjective algebra automorphisms of* A. *Then* ϕ *can be uniquely represented in the form*

$$(\phi f)z = f[\tau(z)] \qquad f\in A, \quad z\in\Delta$$

where $\tau: \Delta \to \Delta$ is a conformal map of Δ onto itself.

Proof: Suppose ϕ is an algebra automorphism of A. Let $f\in A$ and $\lambda\in\mathbb{C}$ be given. It is clear that λ belongs to the range $f(\bar{\Delta})$ of f if and only if $\lambda e-f$ is not invertible in A. Since ϕ is an algebra automorphism, $\lambda e-f$ is invertible if and only if $\phi(\lambda e-f) = \lambda e-\phi f$ is invertible; thus f and ϕf have the same range. Applying this to the identity function $id_\Delta\in A$ we get that $\tau =: \phi\, id_\Delta$ is continuous on $\bar{\Delta}$, holomorphic on Δ and $\tau(\bar{\Delta}) = \bar{\Delta}$. In particular, τ is not constant and, as it is holomorphic, $\tau(\Delta)$ is an open set. Thus $\tau(\Delta)\subset\Delta$. Next let $z\in\Delta$ and $f\in A$ be given. Then $\tau(z)\in\Delta$ and we have

$$f-f[\tau(z)] = z[id_\Delta-\tau(z)]g$$

for some $g\in A$. Applying ϕ we obtain

$$\phi f-f[\tau(z)] = (\tau-\tau(z))\phi g$$

so that $\phi f-f[\tau(z)]$ vanishes at z, i.e.

$$(\phi f)z = f[\tau(z)]$$

for all $f\in A$ and $z\in\Delta$. Thus ϕ is the mapping $\phi f = f\circ\tau$. Since ϕ is surjective, by considering its inverse ϕ^{-1} we obtain $\phi^{-1} = f\circ\sigma$, $f\in A$, for some function σ which is continuous on $\bar{\Delta}$, holomorphic on Δ and $\sigma(\Delta)\subset\Delta$. Then we have $f = f\circ(\tau\circ\sigma)$ for all $f\in A$. In particular, for $f = id_\Delta$ we get $\tau\circ\sigma = id_\Delta$, whence $\tau(\Delta) = \Delta$ and τ is a conformal map of the unit disk. #

By a process of compactification, we can also consider algebras of continuous functions on some non compact topological spaces.

<u>8.17. EXERCISES</u>. (a) Let X be a completely regular Hausdorff space and denote by $\mathcal{B}(X)$ the Banach algebra of all bounded continuous functions $f: X \to \mathbb{C}$ endowed with the supremum norm. Show that, if βX denotes the Stone-Čech compactification

of X, then there is a surjective isometric algebra isomorphism $J: \mathcal{B}(X) \to \mathcal{C}(\beta X)$ which commutes with the natural conjugation of $\mathcal{B}(X)$.

(b) Show that, if A is any closed complex subalgebra of $\mathcal{B}(X)$ such that $e \in A$, then the subspace A_0 associated with A by definition 7.5 is given by $A = \{\bar{f} \in A;\ f \in A\}$. Thus

$$(\mathrm{Aut}D)0 = \{f \in A;\ \ \bar{f} \in A,\ \ \|f\| < 1\}$$

and every $F \in \mathrm{Aut}D$ can be represented in the form $F = L \circ M$ where L is a surjective linear isometry of A and M is a Möbius transformation

$$Mf = \frac{f+a}{e+\bar{a}f} \qquad f \in D$$

where $a \in A_0$, $\|a\| < 1$.

Some interesting examples are included in the above situation. For instance

(c) Let $X =: \Delta$ be the open unit disk of $\mathbb{C}$ and denote by $H^\infty(\Delta)$ the algebra of all bounded holomorphic functions $f: \Delta \to \mathbb{C}$. By an argument similar to that of proposition 8.16, show that any surjective algebra automorphism ϕ of $H^\infty(\Delta)$ can be represented in the form $\phi f = f \circ \tau$, $f \in H^\infty$, for some $\tau \in \mathrm{Aut}\Delta$. Thus we have $(\mathrm{Aut}D)0 = \Delta$ and any $F \in \mathrm{Aut}D$ can be written in the form

$$Ff = \alpha\, \frac{f \circ \tau + a}{e + \bar{a}.f \circ \tau} \qquad f \in D$$

where $\tau \in \mathrm{Aut}\Delta$ and α and a are complex numbers with $|\alpha| = 1$, $|a| < 1$.

(d) Assume that X is a discrete topological space. Show that for all $F \in \mathrm{Aut}D$, there exists a unique permutation T of X and a function $M: X \to \mathrm{Aut}\Delta$ such that

$$(Ff)x = M(x)f[T(x)] \qquad x \in X, \quad f \in D.$$

§4.- Operator valued Möbius transformations.

Let H be a fixed complex Hilbert space with scalar product $\langle , \rangle$ and write $\mathcal{B} =: B[\mathcal{L}(H)]$. The transformations defined, for $A \in \mathcal{B}$, analogously as in $\mathbb{C}^1$ by

$$M'(X) =: (X+A)(1+A^*X)^{-1} \quad , \quad M''(X) =: (1+XA^*)^{-1}(X+A), \qquad X \in \mathcal{B}$$

seem to be the natural candidates to non-linear elements of Aut$\mathcal{B}$. Unlike in the case of $\mathbb{C}^1$, the situation $M'_{A^*}(X^*) \neq M'_A(X)^*$ may occur; therefore, in general, these transformations cannot belong to Aut$\mathcal{B}$. However, as it was observed by Potapov |48|, a slight correction transforms them into elements of Aut$\mathcal{B}$ as we shall see.

8.18 LEMMA. *Given* $X, Y \in \mathcal{L}(H)$ *and* $G \in GL(H)$, *we have* $X \geq Y$ *if and only if* $G^*XG \geq G^*YG$.

Proof: $X \geq Y \Leftrightarrow \langle (X-Y)f, f\rangle \geq 0 \quad \forall f \in H \Leftrightarrow$
$\Leftrightarrow \langle (X-Y)Gf, Gf\rangle \geq 0 \quad \forall f \in H \Leftrightarrow \langle G^*(X-Y)Gf, f\rangle \geq 0 \ , \quad \forall f \in H.$

#

8.19. THEOREM. *For each* $A \in \mathcal{B}$, *the mapping*

$$M_A(X) =: (1-AA^*)^{-\frac{1}{2}}(X+A)(1+A^*X)^{-1}(1-A^*A)^{\frac{1}{2}} \ , \quad X \in \mathcal{B}$$

is a biholomorphic automorphism of $\mathcal{B}$ *and we have* $M_A^{-1} = M_{-A}$.

Proof: Consider the power series of M'_A

$$M'_A(X) = (X+A)\sum_{n=0}^{\infty}(-A^*X)^n = A + \sum_{n=0}^{\infty}(-1)^n(1-AA^*)\underbrace{XA^*XA^*\ldots A^*X}_{(2n+1)\text{ terms}}$$

Similarly

$$M''_A(X) = A + \sum_{n=0}^{\infty} (-1)^n \underbrace{XA^*XA^* \ \dots \ A^*X}_{(2n+1)\ \text{terms}}(1-A^*A)$$

Therefore,

$$M'_A(X)(1-A^*A) = (1-AA^*)M''_A(X) \qquad X \in \mathcal{B}$$

that is,

$$M_A(X) = (1-AA^*)^{-\frac{1}{2}}M'_A(X)(1-A^*A)^{\frac{1}{2}} = (1-AA^*)^{\frac{1}{2}}M''_A(X)(1-A^*A)^{-\frac{1}{2}}$$

Since

$$M'_A(X)^* = (1+X^*A)^{-1}(X^*+A^*) = M''_{A^*}(X^*),$$

it immediately follows that

$$M_A(X)^* = M_{A^*}(X^*)$$

holds for all $X \in \mathcal{B}$. Now we can show that $M_A(\mathcal{B}) \subset \overline{\mathcal{B}}$, i.e.

$$(8.16) \qquad \| M_A(X) \|^2 \leqslant 1 \qquad \text{or} \qquad M_A(X)^*M_A(X) \leqslant 1 \qquad X \in \mathcal{B}$$

Indeed, observe that

$$M_A(X)^*M_A(X) = M_{A^*}(X^*)M_A(X) =$$

$$= (1-A^*A)^{\frac{1}{2}}M''_{A^*}(X^*)(1-AA^*)^{-\frac{1}{2}}(1-AA^*)^{-\frac{1}{2}}M'_a(X)(1-A^*A)^{\frac{1}{2}}$$

Thus, by applying the previous lemma with
$G =: (1-A^*A)^{-\frac{1}{2}}(1+A^*X)$, (8.16) is equivalent to

$$(8.17) \qquad (X^*+A^*)(1-AA^*)^{-1}(X+A) \leqslant (1+X^*A)(1-AA^*)^{-1}(1+A^*X)$$

Since

$$X^*A(1-A^*A)^{-1}A^*X-X^*(1-AA^*)^{-1}X= -X^*X$$

$$(1-A^*A)^{-1}-A^*(1-AA^*)^{-1}A= 1$$

$$X^*A(1-A^*A)^{-1}-X^*(1-AA^*)^{-1}A= 0= (1-A^*A)^{-1}A^*X-A^*(1-AA^*)^{-1}X$$

as it can be seen from the corresponding power series expansions, the difference between the right hand side and the left hand side in (8.17) equals $1-X^*X$ which is a positive operator whenever $\|X\|<1$. Thus (8.16) is established.

Clearly, the Potapov-Möbius transformation M_A is holomorphic. So it remains to prove that $M_A \circ M_{-A}= \mathrm{id}$, and it suffices to see that $M_A[M_{-A}(Y)]= Y$ for all Y in some neighbourhood of $-A$. Now, the relation $M_A(X)= Y$ is equivalent to

$$M'_A(X)= Y' \qquad \text{where} \qquad Y'=: (1-AA^*)^{\frac{1}{2}}Y(1-A^*A)^{-\frac{1}{2}}$$

i.e. $\quad X+A= Y'(1+A^*X), \quad$ or $\quad X= (1-Y'A^*)^{-1}(Y'-A)= M''_{-A}(Y'),$

that is,

$$M_A^{-1}(Y)= M''_A[(1-AA^*)^{\frac{1}{2}}Y(1-A^*A)^{-\frac{1}{2}}]$$

whenever $\|Y'\|<1$. But now

$$Y'A^*= (1-AA^*)^{\frac{1}{2}}Y(1-A^*A)^{-\frac{1}{2}}A^*=$$

$$= (1-AA^*)^{\frac{1}{2}}Y \sum_{n=0}^{\infty} \binom{-\frac{1}{2}}{n} \underbrace{A^*AA^*A \ldots AA^*}_{2n+1 \text{ terms}} = (1-AA^*)^{\frac{1}{2}}YA^*(1-AA^*)^{-\frac{1}{2}}$$

and hence

$$(1-Y'A^*)^{-1}= \sum_{n=0}^{\infty} (1-AA^*)^{\frac{1}{2}}(YA^*)^n(1-AA^*)^{-\frac{1}{2}}=$$

$$= (1-AA^*)^{\frac{1}{2}}(1-YA^*)^{-1}(1-AA^*)^{-\frac{1}{2}}$$

Therefore

$$M''_{-A}(Y') = (1-AA^*)^{\frac{1}{2}}(1-YA^*)^{-1}(1-AA^*)^{-\frac{1}{2}}\left[(1-AA^*)^{\frac{1}{2}}Y(1-A^*A)^{-\frac{1}{2}}-A\right] =$$

$$= (1-AA^*)^{\frac{1}{2}}(1-YA^*)^{-1}(Y-A)(1-A^*A)^{-\frac{1}{2}} = M_{-A}(Y)$$

#

8.20. COROLLARY. *The domain B is homogeneous; namely* $\{M_A(0);\ A\in B\} = B$.

8.21. LEMMA. *Let $A\in B$ be fixed. Then for all $X\in B$ and $Z\in L(H)$, we have*

$$(M_A)_X^{(1}Z = (1-AA^*)^{\frac{1}{2}}(1+XA^*)^{-1}Z(1+A^*X)^{-1}(1-A^*A)^{\frac{1}{2}}$$

Proof: We know that

$$M'(X) = A+(1-AA^*)\sum_{n=0}^{\infty}(-1)^n \underbrace{XA^*XA^* \ldots A^*X}_{(2n+1)\text{ terms}} = A+(1-AA^*)X(1+A^*X)^{-1}$$

Hence

$$(M'_A)_X^{(1}Z = (1-AA^*)Z(1+A^*X)^{-1} - (1-AA^*)X(1+A^*X)^{-1}A^*Z(1+A^*X)^{-1} =$$

$$= (1-AA^*)\left[1-X(1+A^*X)^{-1}A^*\right]Z(1+A^*X)^{-1}$$

Since

$$1-X(1+A^*X)^{-1}A^* = 1-XA^*+XA^*XA^*+\ldots+(-1)^n(XA^*)^n+\ldots+ = (1+XA^*)^{-1}$$

and since

$$(M_A)_X^{(1}Z = (1-AA^*)^{-\frac{1}{2}}\left[(M'_A)_X^{(1}Z\right](1-A^*A)^{\frac{1}{2}}$$

the proof is complete.

#

8.22. EXERCISE. Show that, if $\psi: \Delta\to\mathbb{C}$ is holomorphic and $\|A\|<1$, then $\psi(A^*A)A^* = A^*\psi(AA^*)$.

Next we are going to study the one-parameter groups of Potapov-Möbius transformations.

8.23. PROPOSITION. *Let* A, B∈*B be such that both* AB^* *and* A^*B *are selfadjoint operators. Then* $M_A \circ M_B = M_{M_A(B)}$.

Proof: By Cartan's uniqueness theorem, it suffices to show that the automorphisms $M_A \circ M_B$ and $M_{M_A(B)}$ have the same image and the same derivative at the origin. Obviously $M_A[M_B(0)] = M_A(B) = M_{M_A(B)}(0)$.

On the other hand, from lemma 8.21 we get

$$(M_{M_A(B)})_0^{(1}Z = [1-M_A(B)M_A(B)^*]^{\frac{1}{2}} Z [1-M_A(B)^* M_A(B)]^{\frac{1}{2}}$$

In the course of the proof of (8.16) we have established that

$$G^*[1-M_A(X)^*M_A(X)]G = 1-X^*X \quad \text{where} \quad G =: (1-A^*A)^{-\frac{1}{2}}(1+A^*X)$$

and hence

$$1-M_A(B)^*M_A(B) = (1-A^*A)^{\frac{1}{2}}(1+B^*A)^{-1}(1-B^*B)(1+A^*B)^{-1}(1-A^*A)^{\frac{1}{2}}$$

Observe that, from our assumption, the operators B^*B, $B^*A = A^*B$ and A^*A commute. Therefore

$$[1-M_A(B)^*M_A(B)]^{\frac{1}{2}} = (1-A^*A)^{\frac{1}{2}}(1+A^*B)^{-2}(1-B^*B)^{\frac{1}{2}}$$

$$[1-M_A(B)M_A(B)^*]^{\frac{1}{2}} = [1-M_{A^*}(B^*)^*M_{A^*}(B^*)]^{\frac{1}{2}} = (1-AA^*)^{\frac{1}{2}}(1+AB^*)^{-1}(1-BB^*)^{\frac{1}{2}}$$

Thus

$$(M_{M_A(B)})_0^{(1}Z =$$

$$= (1-AA^*)^{\frac{1}{2}}(1+BA^*)(1+BA^*)^{-1}(1-BB^*)^{\frac{1}{2}}Z(1-B^*B)^{\frac{1}{2}}(1+AB^*)^{-1}(1-A^*A)^{\frac{1}{2}}$$

On the other hand, by the chain rule and lemma 8.21, we also have

$$(M_A \circ M_B)_0^{(1} Z = (M_A)_B^{(1} (M_B)_0^{(1} Z =$$

$$= (1-AA^*)^{\frac{1}{2}} (1+BA^*)^{-1} (1-BB^*)^{\frac{1}{2}} Z (1-B^*B)^{\frac{1}{2}} (1+AB^*)(1-A^*A)^{\frac{1}{2}}.$$

#

8.24. THEOREM. *Let* $t \to M_{A_t}$ *be a continuous one-parameter group of Potapov-Möbius transformations. Then the operator* $A_t^* A_s$ *is selfadjoint for all* $t,s \in \mathbb{R}$ *if and only if, for some* $D \in L(H)$, *we have*

$$M_{A_t} = \exp t(D - {}^sD^*) \qquad t \in \mathbb{R}$$

where $D-{}^sD^*$ *stands for the vector field* $X \to D - XD^*X$, $X \in L(H)$.

Proof: Let $t \to M_{A_t}$ be a continuous one-parameter group. Then

$$M_{A_t} = \exp tV \qquad t \in \mathbb{R}$$

holds for some $V \in \operatorname{aut} B$, so that

$$V(X) = \frac{d}{dt}\Big|_0 M_{A_t}(X) \qquad X \in B$$

Thus, by setting $D =: \frac{d}{dt}\Big|_0 M_{A_t}(0) = \frac{d}{dt}\Big|_0 A_t$, we have $A_t = tD + \omega(t)$ where $\lim_{t\to 0} \frac{1}{t}\omega(t) = 0$, and

$$V(X) = \frac{d}{dt}\Big|_0 M_{A_t}(X) = \frac{d}{dt}\Big|_0 M_{tD+\omega(t)}(X) = \frac{d}{dt}\Big|_0 M_{tD}(X) =$$

$$= (M_A)_0^{(1} D = \frac{d}{dt}\Big|_0 (1-t^2DD^*)^{-\frac{1}{2}} (X+tD)(1+tD^*X)^{-1} (1-t^2D^*D)^{\frac{1}{2}} =$$

$$= D - XD^*X$$

Conversely, we shall prove that, given $D \in L(H)$, the mapping $t \to \exp t(D-{}^sD^*)$, $t \in \mathbb{R}$, is a group of Potapov-Möbius transformations.

Let $D = J|D|$ be the *polar decomposition* of D, i.e., $|D| =: (D^*D)^{\frac{1}{2}}$ and J is a suitable isometry of the (closure of the) range of D into H. Remark that $D^n = J|D|^n$ for $n\in\mathbb{N}$. Since the function *tanh* (hyperbolic tangent) is the solution of the differential equation

$$\frac{dy}{dt} = 1-y^2(t) \qquad y(0) = 0$$

and $tanh(t)\in[0,1)$ for $t\in\mathbb{R}$, the function $\mathbb{R}\to\mathcal{L}(H)$ given by $A_t =: J\,tanh(t|D|)$ satisfies

$$\frac{d}{dt} A_t = D - A_t D^* A_t, \qquad A_o = 0, \qquad A_t \in \mathcal{B}$$

for all $t\in\mathbb{R}$. Moreover,

$$A_t^* A_s = tanh(t|D|)\circ tanh(s|D|), \qquad A_s A_t^* = J\circ tanh(t|D|)\circ tanh(s|D|)^*$$

for $s,t\in\mathbb{R}$. Thus, by propostion 8.23,

$$M_{A_t} \circ M_{A_s} = M_{M_{A_t}(A_s)} \qquad s,t\in\mathbb{R}$$

Therefore, it suffices to prove that

$$(8.18) \qquad M_{A_t}(A_s) = A_{t+s} \qquad s,t\in\mathbb{R}$$

because (8.18) implies that $t\to M_{A_t}$ is a one-parameter group of Potapov-Möbius transformations such that $\frac{d}{dt}\Big|_0 M_{A_t}(0) = D$ and therefore

$$M_{A_t} = \exp t(D - {}^sD^*) \qquad t\in\mathbb{R}$$

Now we have

$$M_{A_t}(A\ ) = (1-A_t A_t^*)^{-\frac{1}{2}}(A_s + A_t)(1+A_t^* A_s)^{-1}(1-A_t^* A_t)^{\frac{1}{2}} =$$

$$=(1-A_tA_t^*)^{-\frac{1}{2}}J\left[tanh(s|D|)+tanh(t|D|)\right]\circ\left[1+tanh(t|D|)\,tanh(s|D|)\right]^{-1}\circ$$

$$\circ\left[1-tanh^2(t|D|)\right]^{\frac{1}{2}}=$$

$$=(1-A_tA_t^*)^{-\frac{1}{2}}J\left[tanh(s+t)|D|\right]\circ\left[1-tanh^2(t|D|)\right]^{\frac{1}{2}}=$$

$$=(1-A_tA_t^*)^{-\frac{1}{2}}J\left[1-tanh(t|D|)^2\right]^{\frac{1}{2}}tanh(s+t)|D|$$

Here

$$(1-A_tA_t^*)^{-\frac{1}{2}}J=J-\binom{\frac{1}{2}}{n}A_tA_t^*+\binom{\frac{1}{2}}{n}A_tA_t^*A_tA_t^*-\ldots=$$

$$=J-\binom{\frac{1}{2}}{1}Jtanh^2(t|D|)+\binom{\frac{1}{2}}{2}Jtanh^4(t|D|-\ldots=$$

$$=J\left[1-tanh^2(t|D|\right]^{-\frac{1}{2}}$$

Hence it follows that

$$M_{A_t}(A_s)=J\left[1-tanh^2(t|D|)\right]^{-\frac{1}{2}}\left[1-tanh^2(t|D|)\right]^{\frac{1}{2}}tanh(s+t)|D|=$$

$$=Jtanh(s+t)|D|=A_{s+t}\ .$$

#

<u>8.25. COROLLARY</u>. *We have*

$$\text{aut}\mathcal{B}=\{D-{}^{s}D^*;\ \ D\in\mathcal{L}(H)\}$$

§5.- <u>J^*-algebras of operators.</u>

In view of theorems 7.28 and 8.19, in order to achieve the complete description of the elements of Aut$\mathcal{B}$ we need only to compute the subgroup Aut$^0\mathcal{B}$ which can be identified with the family of all surjective linear isometries of $\mathcal{L}(H)$. To solve this problem, it is a natural approach to look for certain elements of $\mathcal{L}(H)$ with rather specific geometrical properties that must be preserved by all surjective linear isometries of $\mathcal{L}(H)$. Namely, one conjectures that the operators

$$u\otimes v^*: h\to\langle h,v^*\rangle u, \qquad h\in H$$

with $u,v\in H$, meet this requirement. However, a direct geometrical characterization of them is rather sophisticated. On the other hand, their algebraic description is quite convenient as Corollary 8.25 enables us to extend the use of Potapov-Möbius transformations in a far reaching algebraic direction.

8.26. DEFINITION. *A closed complex subspace* E *of* $L(H)$ *is said to be a* J^**-algebra on* H *if* $XA^*X\in E$ *whenever* $X,A\in E$. *We endow* E *with the "trilinear product"*

$$2(X,A^*,Y)=: XA^*Y+YA^*X$$

With respect to this product, E is a *ternary algebra*. Indeed, for $X,A,Y\in E$, we have

$$2(X,A^*,Y)= (X+Y)A^*(X+Y)-XA^*X-YA^*Y\in E$$

A linear mapping $L: E\to E$ *is a* J^**-automorphism of* E *if* $L(E)=E$ *and*

$$L(X,A^*,Y)= (LX, (LA)^*, LY)$$

holds for all $X,A,Y\in E$.

A J^**-derivation of* E *is a linear mapping* $L: E\to E$ *such that*

$$L(X,A^*,Y)= (LX,A^*,Y)+(X,(LA)^*,Y)+(X,A^*,LY)$$

holds for all $X,A,Y\in E$.

Many familiar spaces are J^*-algebras. For instance, every C^*-algebra is obviously a J^*-algebra with respect to the natural triple product. Thus, by the Gel'fand-Naimark theorem, every B^*-algebra is a J^*-algebra, too. Also any Hilbert space H can be identified with $\{x\otimes e^*;\ x\in H\}$ where e is any fixed unit vector of H; therefore H is a J^*-algebra, too.

Cartan factors, which can be considered as the natural infinite dimensional generalizations of the spaces appearing in Cartan's classification of bounded symmetric domains (cf. Chap. 9), are of particular interest among the J^*-algebras on H. To define them, let us recall that a *conjugation* on H is a conjugate linear mapping $Q: H \to H$ such that $\|Q\| = 1$ and $Q^2 = 1$. Given any conjugation Q on H, we can find a complete orthonormal system $S = \{e_j;\ j \in J\}$ in H such that

$$Qe_j = e_j \qquad Q(ie_j) = -ie_j \qquad j \in J$$

Thus

$$\langle Qx, Qy\rangle = \langle y, x\rangle \qquad x, y \in H$$

and, with respect to S, the matrix of the operator QA^*Q is the transposed of that of A, i.e.

$$\langle QA^*Ae_j,\ e_k\rangle = \langle Ae_k, e_j\rangle, \qquad j,k \in J$$

The operation $A \to A^\tau =: QA^*Q$, $A \in L(H)$, is called the *transposition* associated with Q.

<u>8.27. DEFINITION</u>. *A complex subspace F of* $L(H)$ *is said to be a Cartan factor of:*

type I *if* $F = \{X \in L(H);\ X = P_2XP_1\}$ *for some fixed orthogonal projectors* P_1, P_2.

type II *if* $F = \{X \in L(H);\ X^\tau = X\}$ *for some transposition on* H.

type III *if* $F = \{X \in L(H);\ X^\tau = -X\}$ *for some transposition on* H.

type IV *if F is a closed complex subspace of* $L(H)$ *such that* $\{X^2;\ X \in F\} \subset \mathbb{C}.1$ *and* $\{X^*;\ X \in F\} \subset F$.

<u>8.28. EXERCISE</u>. Verify that Cartan factors are J^*-algebras.

8.29. LEMMA. (a) *A closed complex subspace* E *of* $L(H)$ *is a* J^**-algebra if and only if* $AA^*A \in E$ *whenever* $A \in E$.

(b) *Given a* J^**-algebra* E *on* H, *a map* $L \in L(E)$ *is a* J^**-isomorphism of* E *if and only if* $L(E) = E$ *and* $L(AA^*A) = LA(LA)^*LA$ *holds for all* $A \in E$.

(c) *A map* $L \in L(E)$ *is a* J^**-derivation on* E *if and only if we have*

$$L(AA^*A) = (LA)A^*A + A(LA)^*A + AA^*(LA), \qquad A \in E$$

Proof: The statements are immediate consequences of the fact that, for $X, A, Y \in L(H)$,

$$(X,A^*,Y) \in \mathrm{Span}\{(X+\varepsilon_1 A+\varepsilon_2 Y)(X+\varepsilon_1 A+\varepsilon_2 Y)^*(X+\varepsilon_1 A+\varepsilon_2 Y);\ \varepsilon_1^4,\ \varepsilon_2^4 = 1\}$$

8.30. THEOREM. *Let* E *denote a* J^**-algebra on* H. *Then every Potapov-Möbius transformations* M_A *with* $A \in B(E)$ *maps* $B(E)$ *onto itself and*

$$\mathrm{Aut}\, B(E) = \{M_A \circ L;\ A \in B(E),\ \ L \text{ is a } J^*\text{-automorphism of } E\}$$

$$\mathrm{aut}\, B(E) = \{A + L - {}^sA^*;\ A \in E,\ L \text{ is a } J^*\text{-derivation of } E\}$$

where $A+L-{}^sA^*$ *denotes the vector field* $X \to A + L(X) - XA^*X$, $X \in B(E)$.

Proof: Let $A \in B(E)$ be arbitrarily fixed and let $A = J|A|$ be its polar decomposition. By setting $D =: \sum_{n=0}^{\infty} \frac{1}{2n+1} (AA^*)^n A$, we have

$$D = J \sum_{n=0}^{\infty} \frac{1}{2n+1} |A|^{2n+1} = J\, \tanh^{-1}|A|$$

Therefore $A = J\, \tanh|D|$ and so $M_A = \exp(D - {}^sD^*)$. From the power series expansion we see that $D \in E$, whence it follows that $D - XD^*X \in E$, whenever $X \in E$. Thus, M_A maps $B(E)$ into itself.

Similarly, the inverse M_{-A} of M_A maps $B(E)$ into itself, so that $M_A[B(E)] = B(E)$.

Since $\{M_A(0);\ A\in B(E)\} = B(E)$ and since $\{A - {}^sA^*;\ A\in E\} = aut_0 B(E)$, it suffices to see that, given $L\in L(E)$, we have

(8.19) $L\in Aut^0 B(E) \Leftrightarrow L$ is a J^*-automorphism of E

(8.20) $L\in aut^0 B(E) \Leftrightarrow L$ is a J^*-derivation on E

Proof of (8.19): Suppose that $L\in Aut^0 B(E)$. From proposition 7.9(d) we see that

$$X(LA)^*X = L[(L^{-1}X)A^*(L^{-1}X)], \qquad X, A\in E$$

Thus $(LA)(LA)^*(LA) = L(AA^*A)$ for all $A\in E$.

Conversely, suppose that L is a J^*-automorphism. Then $L(E) = E$. Moreover, for $X\in L(H)$ with $X = J|X|$, we have $XX^*X = J|X|^3$; therefore

$$\|XX^*X\| = \||X|^3\| = \rho(|X|^3) = \rho^3(|X|) = \|X\|^3,$$

so that

$$\|LA\|^3 = \|(LA)(LA)^*(LA)\| = \|L(AA^*A)\| \leqslant \|L\|\ \|AA^*A\| = \|L\|\ \|A\|^3$$

holds for all $A\in E$. Thus $\|L\| \leqslant 1$. But L^{-1} is also a J^*-automorphism of E and, by the open mapping theorem $L^{-1}\in L(E)$. Thus $\|L^{-1}\| \leqslant 1$ and L is a surjective linear isometry of E.

Proof of (8.20). Suppose that $L\in aut^0 B(E)$, and define $G^t =: \exp tL$ for $t\in\mathbb{R}$. Then $G^t\in Aut^0 B(E)$ and so, for each fixed $A\in E$, we have

$$L(AA^*A) = \frac{d}{dt}\Big|_0 G^t(AA^*A) = \frac{d}{dt}\Big|_0 (G^tA)(G^tA)^*(G^tA) =$$

$$= \left(\frac{d}{dt}\Big|_0 G^tA\right)A^*A + A\left(\frac{d}{dt}\Big|_0 G^tA\right)^*A + AA^*\left(\frac{d}{dt}\Big|_0 G^tA\right) =$$

$$= (LA)A^*A + A(LA)^*A + AA^*(LA)$$

Conversely, let L be a J^*-derivation on E and set again $G^t =: \exp tL$ for $t \in \mathbb{R}$. We must show that, for $t \in \mathbb{R}$, $G^t \in \mathrm{Aut}^0 B(E)$, or equivalently, that G^t is a J^*-automorphism of E. For fixed $A \in E$, we have

$$\frac{d}{dt} G^{-t}(G^tA, (G^tA)^*, G^tA) = -LG^t(G^tA, (G^tA)^*, G^tA) +$$

$$+G^{-t}(LG^tA, (G^tA)^*, G^tA) + G^{-t}(G^tA, (LG^tA)^*, G^tA) +$$

$$+G^{-t}(G^tA, (G^tA)^*, LG^tA) =$$

$$= -G^{-t}L(G^tA, (G^tA)^*, G^tA) + G^{-t}L(G^tA, (G^tA)^*, G^tA) = 0$$

for all $A \in E$ and $t \in \mathbb{R}$. Thus

$$G^{-t}(G^tA), (G^tA)^*, G^tA) = G^0(G^0A, (G^0A)^*, G^0A) = (A, A^*, A)$$

whence $G^t \in \mathrm{Aut}^0 B(E)$.

#

<u>8.31. EXERCISE</u>. Let E_1 and E_2 be J^*-algebras on H and suppose that $L \in \mathcal{L}(E_1, E_2)$ is a bijective mapping. Then L is isometric if and only if it is a J^*-isomorphism.

§6.- <u>Minimal partial isometries in Cartan factors</u>.

Recall that an operator $J \in \mathcal{L}(H)$ is called a *partial isometry* if, for some subspace H_0 of H, the restriction $J|_{H_0}$ is an isometry and $J|_{H_0^\perp} = 0$. It is a well known consequence of the existence of the polar decomposition that

(8.21) A is a partial isometry $\Leftrightarrow AA^*A = A$, $A \in L(H)$

From (8.21) we can easily obtain the following

8.32. LEMMA. *Let* E *be a* J^**-algebra on* H. *Then any linear automorphism of* $B(E)$ *preserves the set of partial isometries of* E .

8.33. DEFINITION. *Let* E *be a* J^**-algebra and assume that* $J_1, J_2 \in E$ *are partial isometries. Let us set*
$H_k =: \{x \in H; \ \|J_k x\| = \|x\|\}$ *for* $k = 1,2$. *We say that* J_1 *is a part of* J_2 *if* $H_1 \subset H_2$ *and* $J_{2|H_1} = J_{1|H_1}$. *We write*
$J_1 < J_2$ *if* J_1 *is part of* J_2.
Clearly, the relation < is a partial ordering on the set of non-zero partial isometries of E . The minimal elements with respect to < are called *minimal partial isometries* of E .

We recall that a net $(A_j)_{j \in J}$ in $L(H)$ is said to be convergent to A with respect to the weak operator topology if we have $\lim_j \langle A_j x, y\rangle = \langle Ax, y\rangle$ for all $x, y \in H$. We write T_w for the weak operator topology on $L(H)$.

8.34. THEOREM. *If the* J^**-algebra* E *is closed in* $L(H)$ *with respect to* T_w, *then* E *is the* T_w*-closure of the linear hull of* $p(E)$ *and*

$$mp(E) = \{A \in E \ ; \ AA^*A = A \neq 0, \quad A E^* A = \mathbb{C}A\}$$

Proof: Let $A \in E$ be arbitrarily fixed and suppose that

$$A = J|A| = J\int_{[0, \|A\|]} \lambda dP(\lambda),$$

where $\lambda \to P(\lambda)$ is the spectral measure of $|A|$, is the polar decomposition of A (cf. |13|).

Consider the sequence $Y_n =: (AA^*)^n A$, $n = 0,1,\ldots$ Since
$Y_{n+1} = AY_n^* A$,

$$Y_n = J \int_{[0,\|A\|]} \lambda^{2n+1} dP(\lambda) \in E$$

holds for all $n \in \mathbb{N}$. Therefore, if the sequence of odd polynomials $(p_n)_{n \in \mathbb{N}}$ is bounded on $[0,\|A\|]$ and p_n converges pointwise on $[0,\|A\|]$ to some function ψ, we have

$$J \int_{[0,\|A\|]} \psi(\lambda) dP(\lambda) \in E$$

In particular, $JP[\alpha,\beta] \in E$ for each $\alpha,\beta \in \mathbb{R}_+$. However, it is well known that each operator $JP[\alpha,\beta]$, $\alpha,\beta \in \mathbb{R}$, is a partial isometry, and that

$$\{ \int_{[0,\|A\|]} \psi(\lambda) dP(\lambda) ;\ \psi \text{ Borel function}\} = \overline{\text{Span}\{P[\alpha,\beta];\ \alpha,\beta \in \mathbb{R}_+\}}^{T_w}$$

Therefore A belongs the T_w-closure of Span(E).

Suppose now that $\tilde{J} \in \text{mp}(E)$ and let $X \in E$ be given. Define A by $A =: \tilde{J}X^*\tilde{J}$. Clearly $A \in E$ and $\ker\tilde{J} \subset \ker A$, whence

$$\text{range}|A| \subset (\ker A)^{\perp} \subset (\ker\tilde{J})^{\perp} = \{x \in H;\ \|\tilde{J}x\| = \|x\|\}$$

Thus, if the polar decomposition of A is again $A = J|A|$, then

$$\{x \in H;\ \|Jx\| = \|x\|\} = (\ker A)^{\perp} \subset \{x \in H;\ \|\tilde{J}x\| = \|x\|\}.$$

Therefore we have $J < \gamma\tilde{J}$ for some $\gamma \in \mathbb{C}$, $|\gamma| = 1$, and, as $\tilde{J}$ is minimal, $J = \gamma\tilde{J}$. But now for any spectral projection $P[\alpha,\beta]$ of $|A|$, the operator $JP\{\alpha,\beta]$ is a partial isometry contained in $J = \gamma\tilde{J}$. It follows that $JP(\{\|A\|\}) = J$. Therefore

$$A = J|A| = J \int_{[0,\|A\|]} \lambda dP(\lambda) = JP(\{\|A\|\}) \int_{[0,\|A\|]} \lambda dP(\lambda) =$$

$$= \|A\|\, JP(\{\|A\|\}) = \|A\|\, J = \gamma \|A\| \tilde{J}$$

#

8.35. PROPOSITION. *Let* E *be any* J^**-algebra which is* T_w*-closed and assume that for any* $J \in p(E)$ *there is some* $\tilde{J} \in mp(E)$ *such that* $\tilde{J} < J$. *Then* $E = \overline{(\mathrm{Span}\; mp(E))}^{T_w}$.

Proof: Let F be the family of the finite sums $\tilde{J}_1 + \ldots + \tilde{J}_n$, $n \in \mathbb{N}$, of mutually orthogonal minimal partial isometries $\tilde{J}_k \in mp(E)$, $1 \leq k \leq n$, i.e.

$$\{x \in H;\ \|\tilde{J}_k x\| = \|x\|\} \perp \{x \in H;\ \|\tilde{J}_\ell x\| = \|x\|\}$$

for all pairs k, ℓ with $k \neq \ell$. Then, the linear hull of F is T_w-dense in E because, given any $J \in p(E)$, the net $F(J) =: \{\tilde{J} \in F;\ \tilde{J} < J\}$ is not empty and it is weakly convergent to J.

#

8.36. COROLLARY. *In particular, if every* $J \in p(E)$ *contains some* $\tilde{J} \in mp(E)$ *such that* dim range $\tilde{J} < \infty$ *then* $E = \overline{(\mathrm{Span}\; mp(E))}^{T_w}$.

8.37. PROPOSITION. *All Cartan factors of* $L(H)$ *are* T_w*-closed.*

Proof: Given any operators R_1, $R_1 \in L(H)$ and a conjugation Q on H, the mappings $L(H) \to L(H)$ given by

$$L_{R_1 R_2} : X \to R_1 X R_2 \quad , \quad T_Q : X \to QX^*Q$$

are T_w-continuous. On the other hand, if F_k is any Cartan factor of type k, $1 \leq k \leq 3$, we have

$$F_1 = \{X \in L(H);\ L_{P_1 P_2} X = X\} \quad \text{for some projectors } P_1, P_2$$

$$F_2 = \{X \in L(H);\ T_Q X = X\} \quad , \quad F_3 = \{X \in L(H);\ T_Q X = -X\}$$

As for Cartan factors F_4, the proof can be found in |19|, page 334.

8.38. DEFINITION. *Given a* J^**-algebra* E *on* H, *we say that an operator* $A\in E$ *is minimal if* $AE^*A=\mathbb{C}A$. *We write* $m(E)$ *for the set of minimal elements of* E.

It is immediate that the set $m(E)$ is preserved by all J^*-isomorphisms L of E. Moreover, if E is T_w-closed, then

$$mp(E)=m(E)\cap p(E)$$

8.39. THEOREM. *Let* P_1, P_2 *and* Q *be, respectively, orthogonal projectors with* $H_j =:$ range P_j (j= 1,2) *and a conjugation on* H. *Denote by*

$$F_1=\{X\in L(H);\ P_2XP_1=X\},\qquad F_2=\{X\in L(H);\ QX^*Q=X\}$$

$$F_3=\{X\in L(H);\ QX^*Q=-X\}$$

the corresponding Cartan factors. Furthermore, let F_4 *be any Cartan factor of type IV. Then*

$$m(F_1)=\{f\otimes e^*;\ e\in H_1,\ f\in H_2\},\qquad m(F_2)=\{e\otimes(Qe)^*;\ e\in H\}$$

$$m(F_3)=\{f\otimes(Qe)^*-e\otimes(Qf)^*;\ e,f\in H\}$$

$m(F_4)=\{A\in F_4;\ A^2=0\}$ *provided that* $\dim F_4>1$.

Proof: In general, for $e,f\in H$ with $e\neq 0\neq f$, the two relations

$$\{(f\otimes e^*)A^*(f\otimes e^*);\ A\in L(H)\}=\{\langle A^*f,e\rangle f\otimes e^*;\ A\in L(H)\}\subset\mathbb{C}f\otimes e^*$$

$$(f\otimes e^*)(f\otimes e^*)^*(f\otimes e^*)=\|e\|^2\|f\|^2 f\otimes e^*\neq 0$$

hold. Therefore

$$\{f\otimes e^*;\ f,\ e\in H\}\subset mL(H)$$

Case k= 1. From the above remarks it is easy to check that, for $e\in H_1$ and $f\in H_2$, we have $f\otimes e^*\in m(F_1)$. Conversely, let $A\in m(F_1)$ with $A\neq 0$ be given. Then $A=P_2AP_1$ and we can choose some

$x \in H$ such that $P_2 A P_1 x \neq 0$. In particular, the vectors $e =: P_1 x$ and $f =: P_2 Ae$ satisfy $Ae \neq 0$ and $A^* f \neq 0$ so that

$$(Ae) \otimes (A^* f)^* \neq 0$$

Moreover $f \otimes e^* \in \mathcal{F}_1$, and from the minimality of A we obtain

$$\mathbb{C}A \ni A(f \otimes e^*)^* A = (Ae) \otimes (A^* f)^*$$

whence $A = (\lambda Ae) \otimes (A^* f)^*$ for some $\lambda \in \mathbb{C}$.

Case k= 2. Let $e \in H$ be given. Then $Q[e \otimes (Qe)^*]^* Q = e \otimes (Qe)^*$ so that $e \otimes (Qe)^* \in \mathcal{F}_2$ and, as remarked at the begining of the proof, $e \otimes (Qe)^* \in m(\mathcal{F}_2)$.

Conversely, let $A \in m(\mathcal{F}_2)$ with $A \neq 0$ be given, and choose $e \in H$ such that $Ae \neq 0$. Since $\mathcal{F}_2$ is $*$-invariant, $[e \otimes (Qe)^*]^* \in \mathcal{F}_2$

$$\mathbb{C}A \ni A[e \otimes (Qe)^*]A = (Ae) \otimes (A^* Qe)^*$$

Moreover, from $A \in \mathcal{F}_2$ we derive $A^* Q = QA$, whence $A = f \otimes (Qf)^*$ with $f =: \lambda Ae$ for a suitable $\lambda \in \mathbb{C}$.

Case k= 3. For the sake of shortness, we introduce the notation

$$[v,u] =: v \otimes (Qu)^* - u \otimes (Qv)^* , \qquad v,u \in H$$

Now, let $f,e \in H$ be given; it is easy to see that $Q[f,e]^* Q = -[f,e]$, hence $[f,e] \in \mathcal{F}_3$. For arbitrary $A \in \mathcal{L}(H)$

$$[f,e] A^* [f,e] = M - N$$

holds with

$$M = \langle A^* f, Qe \rangle f \otimes (Qe)^* + \langle A^* e, Qf \rangle e \otimes (Qf)^*$$

$$N = \langle A^* e, Qe \rangle f \otimes (Qf)^* + \langle A^* f, Qf \rangle e \otimes (Qe)^*$$

Since $f \otimes (Qf)^*$ and $e \otimes (Qe)^*$ belong to $\mathcal{F}_2$, it follows that $N \in \mathcal{F}_2$.

Since F_3 is a J^*-algebra, for $A\in F_3$ we have $[f,e]A^*[f,e]\in F_3$. Moreover, from $A\in F_3$ we obtain $A^*Q=-QA$; it follows that

$$M=\langle A^*f,Qe\rangle[f,e]\in F_3$$

Therefore $\quad N=M-[f,e]A^*[f,e]\in F_2\cap F_3=\{0\}$. Thus, for $A\in F_3$

$$(8.22)\qquad [f,e]A^*[f,e]=\langle A^*f,Qe\rangle[f,e]\subset \mathbb{C}[f,e]$$

On the other hand, the relation $[f,e]=0$ holds if and only if f,e are linearly dependent. By putting $A=:[f,e]$ in (8.22) we obtain

$$[f,e][f,e]^*[f,e]=\lambda[f,e]$$

with $\lambda=\|e\|^2\|f\|^2-|\langle e,f\rangle|^2$, whence $[f,e]A^*[f,e]=\mathbb{C}[f,e]$ and $[f,e]\in m(F_3)$.

Conversely, let $A\in m(F_3)$ with $A\neq 0$ be given. For $e,f\in H$ we have

$$\mathbb{C}A\ni A[f,e]^*A=(AQe)\otimes(A^*f)^*-(AQf)\otimes(A^*e)^*$$

Now we can choose e,f such that the vectors AQe and AQf are linearly independent. (Indeed, if we had dim range $A=1$, then A would have the form $A=u\otimes v^*$ for some $u,v\in H$. But $u\otimes v^*\in F_3$ holds if and only if $u\otimes v=0$). Hence

$$\mathbb{C}A\ni f'\otimes e'^*-e'\otimes f'^*=[f',e']$$

for some independent couple $f',e'\in H$ and $A=\lambda[f',e']$.

#

<u>Case k= 4</u>. Suppose that $A\in F_4$ and $A^2=0$. Then, given any $X\in F_4$, we have

$$AX^*A=(AX^*+X^*A)A=[(A+X^*)^2-X^{*2}]A\in\mathbb{C}A$$

i.e., $A\in m(F_4)$. Conversely, let $A\in m(F_4)$ be given; then $AF_4^*A=\mathbb{C}A$ and $A^2=\alpha 1$ for some $\alpha\in\mathbb{C}$. Thus, for some $\beta\in\mathbb{C}$,

$$\beta A = (A^*AA^*)^*A = A^2A^*A^2 = \alpha^2A^*$$

If $\alpha^2 \neq 0$, we have $A^* = \gamma A$ for some $\gamma \in \mathbb{C}$. Since we have assumed that $\dim F_4 > 1$, we may fix $X \in F_4 \setminus \mathbb{C}A$ if $\alpha \neq 0$. But then we get the contradiction

$$\mathbb{C}A \ni A(A^*X^*A^*)^*A = A^2XA^2 = \alpha^2X$$

Thus $\alpha = 0$ and $A^2 = \alpha 1 = 0$.

#

8.40. COROLLARY. *We have*

$$\mathrm{mp}(F_1) = \{f \otimes e^*;\ e \in H_1,\ f \in H_2,\ \|e\| = \|f\| = 1\},$$

$$\mathrm{mp}(F_2) = \{e \otimes (Qe)^*;\ e \in H,\ \|e\| = 1\}.$$

$$\mathrm{mp}(F_3) = \{f \otimes (Qe)^* - e \otimes (Qf)^*;\ e,f \in H,\ e \perp f,\ \|e\| = \|f\| = 1\}$$

Proof: Exercise

8.41. THEOREM. *With the precedent notations,*

(a) *Every element of* $\mathrm{Aut}^0 B(F_k)$, $k = 1,2,3$, *is a continuous operator with respect to the* T_w*-topology on* F_k.

(b) *We have* $F_k = (\mathrm{Span}\ \mathrm{mp}(F_k))^{-T_w}$ *for* $k = 1,2,3$.

Proof: (a) Suppose that $(A_j)_{j \in J}$ is a net such that $T_w\text{-}\lim_j A_j = A$ in F_k, and let $L \in \mathrm{Aut}^0 B(F_k)$ and $f,e \in H$ be arbitrarily fixed. We have to show that

$$\langle L(A_j)f,e\rangle \to \langle L(A)f,e\rangle$$

Case $k = 1$: Write $f' =: P_2f$ and $e' =: P_1e$. Then $f' \otimes e'^* \in m(F_1)$ and we can find vectors $f'' \in H_2$, $e'' \in H_1$, such that

$$L(f' \otimes e'^*) = f'' \otimes e''^* \tag{8.23}$$

For any pair $v \in H_2$, $u \in H_1$, and any operator $X \in F_1$ we have

$$(v\otimes u^*)X^*(v\otimes u^*) = \langle v,Xu\rangle v\otimes u^*$$

Applying this, first to $f'\otimes e'^*$ and A, and then to $f''\otimes e''^*$ and L(A), we obtain

$$(8.24) \qquad (f'\otimes e'^*)A^*(f'\otimes e'^*) = \langle f',Ae'\rangle f'\otimes e'^* \quad ,$$

$$(8.25) \qquad (f''\otimes e''^*)(LA)^*(f''\otimes e''^*) = \langle f'',L(A)e''\rangle f''\otimes e''^*$$

As L is a J^*-automorphism, from (8.23) and (8.24) it follows

$$(f''\otimes e''^*)(LA)^*(f''\otimes e''^*) = \langle f',Ae'\rangle f''\otimes e''^*$$

whence, by compairing with (8.25)

$$\langle f',Ae'\rangle = \langle f'',L(A)e''\rangle$$

In a similar manner we get $\langle f',A e_j'\rangle = \langle f'',L(A_j)e''\rangle$ for all $j\in J$. The result follows by letting $t\to\infty$ and taking into account that $T_w\text{-}\lim_j A_j = A$. The proof for the cases k= 2 and k= 3 is quite similar.

(b) Let $J\in p(F_k)$ be arbitrarily given. It suffices to find some $\tilde{J}\in mp(F_k)$ with dim range $\tilde{J}<\infty$. Let us choose $e\in\{x\in H;\ \|Jx\| = \|x\|\}$ with $\|e\| = 1$. Then, the following choices satisfy the requirement: If k= 1, $J=: (Je)\otimes e^*$. If k= 2, we define $P: H\to H$ to be the projector onto the subspace Span{e,QJe} and put $\tilde{J}=: JP$. (Indeed, then $J(QJe) = Qe$ so that $\{x\in H;\ \|\tilde{J}x\| = \|x\|\}$ and range $\tilde{J}$ are Q-invariant). Finally, for k= 3, we can take again $\tilde{J}=: JP$ where P is as before. Now $J(QJe) = -Qe$.

#

<u>8.42. COROLLARY</u>. *For* k= 1,2,3, *every element* $L\in Aut^0 B(F_k)$ *is uniquely determined by its restriction to the subset* $mp(F_k)$.

#

§7.- Description of $\mathrm{Aut}^0 B(F_1)$ and $\mathrm{aut}^0 B(F_1)$.

Let us set $F_1 =: \{X \in L(H);\ P_2 X P_1 = X\}$ where P_1, P_2 are orthogonal projectors on H, and write $H_j =:$ range P_j $(j = 1,2)$. Furthermore, let L denote any fixed element of $\mathrm{Aut}^0 B(F_1)$.

8.43. LEMMA. *Let* $f \in H_2$, $f \neq 0$, *be given. Then, one of the following statements holds:*

(1_f) *There is a vector* $f' \in H_2$ *and a surjective linear isometry* $U_f : H_1 \to H_1$ *such that*

$$L(f \otimes e^*) = f' \otimes (U_f e)^* \qquad e \in H_1$$

(2_f) *Given a conjugation* Q_2 *on* H_2 , *there is a vector* $f' \in H_2$ *and a surjective linear isometry* $U_f : H_1 \to H_2$ such that

$$L(f \otimes e^*) = (Q_2 U_f e) \otimes f'^* \qquad e \in H_1$$

Proof: Fix any $f \in H_2$, $f \neq 0$, and consider any finite set $\{u_k;\ k = 1,\dots,n\} \subset H_1$. Since $f \otimes u_k^*$ and $X =: f \otimes \sum_1^n u_k^*$ are minimal operators in F_1, we can write

$$L(f \otimes u_k^*) = g_k \otimes v_k^* \quad , \quad L(X) = g \otimes v^*$$

for suitable $g, g_k \in H_2$ and $v, v_k \in H_1$, $1 \leq k \leq n$. Then, as both range $L(X)$ and range$(LX)^*$ are one-dimensional spaces, at least one of the sets $\{g_k\}$, $\{v_k\}$ must consist of mutually parallel vectors (i.e, $g_j \parallel g_k\ \forall_{j,k}$ or $v_j \parallel v_k\ \forall_{j,k}$).

Now we proceed to prove the lemma. Let us fix any pair of independent vectors $e_1, e_2 \in H_1$. (We may assume $\dim H_1 > 1$ as otherwise F_1 would be isomorphic to H_2 whose linear isometries are well known). Write

$$(8.25) \qquad L(f \otimes e_i^*) = g_i \otimes u_i^* \qquad i = 1,2$$

There are two possibilities:

(1_f) The situation $u_1 \| u_2$ does not hold. Then, as previously proved, $g_1 \| g_2$ must hold and by (8.25) we can write

$$L(f \otimes e_i^*) = f' \otimes e_i'^* \qquad i = 1,2$$

for a fixed f' and suitable e_i'. Now, let $e \in H_1$ be arbitrarily given and put $L(f \otimes e^*) = g \otimes u^*$. As $\{e_1', e_2', u\}$ cannot consist of mutually parallel vectors, we must have $g \| f'$ and therefore

$$L(f \otimes e^*) = f' \otimes e'^*$$

holds for a unique $e' \in H_1$. Thus we can define a mapping $U_f: H_1 \to H_1$ by setting

$$L(f \otimes e^*) = f' \otimes (U_f e)^* \qquad e \in H_1$$

It is easy to verify that U_f is a surjective linear isometry of H_1 (because L is a surjective linear isometry of $\mathcal{F}_1$).

(2_f) The situation $u_1 \| u_2$ does hold. The construction of the required elements can be carried out in a similar manner. However, a conjugation Q_2 on H_2 must come up to compensate the fact that the variable factors e, e' of the tensor products $L(f \otimes e^*)$ and $e' \otimes f'$ appear in opposite places.

#

<u>8.44. THEOREM</u>. *Let* $L \in \mathrm{Aut}^0 B(\mathcal{F}_1)$ *be given. Then one of the following statements holds:*

(1) *There are surjective linear isometries* $U: H_1 \to H_1$ *and* $V: H_2 \to H_2$ *such that*

$$L(A) = V P_2 A\, U P_1, \qquad A \in \mathcal{F}_1 \tag{8.26}$$

(2) *Given any conjugations* Q_1, Q_2 *on* H_1 *and* H_2 ,

respectively, there are surjective linear isometries $U: H_1 \to H_2$ *and* $V: H_2 \to H_1$ *such that*

$$(8.27) \qquad L(A) = VQ_1A^*Q_2UP \,, \qquad A \in F_1$$

Proof: Let us fix an arbitrary unit vector $f_0 \in H_2$. Replacing L by $L': A \to Q_1(LA)^*Q_2P_2$ if necessary, we may assume L to satisfy condition (1_{f_0}) in lemma 8.43, i.e.

$$(8.28) \qquad L(f_0 \otimes e^*) = f_0' \otimes (U_{f_0} e)^* \,, \qquad e \in H_1$$

Again, replacing L by $L'': A \to V_0 L(A) U_{f_0} P_1$ if necessary, where $V_0 f_0' = f$, we may assume that

$$(8.29) \qquad L(f_0 \otimes e^*) = f_0 \otimes e^*, \qquad e \in H_1$$

Furthermore, it suffices to establish (8.26) for the operators $A \in F_1$ of the form $A = f \otimes e^*$ where $f \in H_2$, $e \in H_1$.
Now let us fix a unit vector $e_0 \in H_1$, too. Applying lemma 8.43 (interchanging the spaces H_1 and H_2) to the vector e_0, we see that one of the following possibilities holds:

(1_{e_0}) There exist a unit vector $e_0' \in H_1$ and a surjective linear isometry $V_{e_0}: H_2 \to H_2$ such that

$$(8.28') \qquad L(f \otimes e_0^*) = (V_{e_0} f) \otimes e_0'^* \,, \quad f \in H_2$$

(2_{e_0}) Given a conjugation Q_1 on H_1, there exists a vector $e_0' \in H_1$ and a surjective linear isometry $Ve_0: H_2 \to H_1$ such that

$$(8.28'') \qquad L(f \otimes e_0^*) = e_0' \otimes (Q_1 V_{e_0} f)^*, \quad f \in H_2$$

First we show that (8.28") is impossible. Suppose that (8.28") holds. Then we may a ssume $e_0' = f_0$ as it can be seen by applying (8.29) to e_0 and (8.28") to f_0. Now as L is an isometry, we have

$$\| f\otimes e_0^* + f_0 \otimes e^* \| = \| f_0 \otimes (Q_1 V_{e_0} f)^* + f_0 \otimes e^* \|$$

whence it follows

$$\max\{ \| f \| , \| e \| \} = \| Q_1 V_{e_0} f + e \| , \qquad e \perp e_0, \quad f \perp f_0$$

However, the right hand side of this latter equality is a continuously Fréchet real-derivable function of (e,f) whereas the left hand side does not admit a Fréchet real-derivative.

Thus, we necessarily have (8.28'). Replacing L by $L''' : A \to V_{e_0}^{-1} L(A)$ if necessary, we may assume that

$$L(f_0 \otimes e^*) = f_0 \otimes e^* , \quad L(f \otimes e_0^*) = f \otimes e_0^* , \quad e \in H_1, \quad f \in H_2$$

It only remains to prove that $L = id_{\mathcal{F}_1}$. To do this, consider any $e_1 \in H_1$, $f_1 \in H_2$. Since e_0 was arbitrarily fixed, the same argument that was used to establish that (8.28") was impossible shows that, for some surjective linear isometry V_{e_1} (and similarly) U_{f_1} ,

$$L(f_1 \otimes e_1^*) = (V_{e_1} f_1) \otimes e_1^* = f_1 \otimes (U_{f_1} e_1^*)$$

As the ranges and kernels of the tensor products on the right hand sides coincide, it follows that

$$L(f_1 \otimes e_1^*) = \gamma f_1 \otimes e_1^*$$

for some $\gamma \in \mathbb{C}$, $|\gamma| = 1$. Next we show that γ does not depend on e,f. Indeed, if $e_1 \perp e_0$ and $f_1 \perp f_0$, then for some matrix (α_{jk}), $0 \leq j, k \leq 1$, we have

$$\left\| \begin{pmatrix} \alpha_{00} & \alpha_{01} \\ \alpha_{10} & \alpha_{11} \end{pmatrix} \right\| = \left\| \sum_{jk} \alpha_{jk} f_j \otimes e_k^* \right\| = \left\| L \sum_{jk} \alpha_{jk} f_j \otimes e_k^* \right\| =$$

$$= \left\| \begin{pmatrix} \alpha_{00} & \alpha_{01} \\ \alpha_{10} & \gamma\alpha_{11} \end{pmatrix} \right\|$$

Hence, it readily follows that $\gamma = 1$ (cf. exercise below). This fact can be interpreted as

$$L\left(\sum_{jk} \alpha_{jk} f_j \otimes e_k^*\right) = \sum_{jk} \alpha_{jk} f_j \otimes e_k^*$$

whenever $e_1 \perp e_0$ and $f_1 \perp f_0$, $\alpha_{jk} \in \mathbb{C}$, $0 \leq j, k \leq 1$, that is (by the arbitrariness of e_1, f_1), $L = id_{F_1}$.

#

8.45. EXERCISE. Prove the relation $\gamma = 1$. Hint: the matrix $\begin{pmatrix} 1 & 1 \\ 1 & \gamma \end{pmatrix}$ cannot be written as $(\alpha_1\ \alpha_2) \otimes (\alpha_3\ \alpha_4)^*$ if $\gamma \neq 1$.

8.46. EXERCISES. (1) Show that the set of the isometries $L: F_1 \to F_1$ for which (8.26) holds is the identity component of $\mathrm{Aut}^0 B(F_1)$. Describe the other connected components if any.

(2) Let $M \in L(H_2)$ and $N \in L(H_1)$ be selfadjoint operators and write $L_t(A) =: e^{itM} P_2 A e^{itN} P_1$ for $t \in \mathbb{R}$ and $A \in F_1$. Show that the mapping $\mathbb{R} \to \mathrm{Aut}^0 B(F_1)$ given by $t \to L_t$ is a continuous one-parameter group of isometries of F_1 whose associated vector field is

$$f_{M,N}: X \to i(MP_2XP_1 + P_2XNP_1), \qquad X \in F_1$$

With the above notations, show that the vector fields $f_{M,N}$ and $f_{M',N'}$ have the same associated one-parameter group if, and only if, for some $\rho \in \mathbb{R}$ we have

$$M' = M + \rho I, \qquad N' = N - \rho I$$

(3) Let $\mathrm{Her}(H_k) =: \{X \in L(H_k);\ X = X^*\}$ denote the set of hermitian elements of $L(H_k)$ for $k = 1,2$, and endow $i\mathrm{Her}(H_k)$ with its usual Banach-Lie algebra structure. Show that

$$i\,\mathrm{Her}(H_2, H_1) =: \{(iM, iN);\ M \in \mathrm{Her}(H_2),\ N \in \mathrm{Her}(H_1)\}$$

with the product structure and the norm

$$\|(iM, iN)\| = \max\{\|M\|, \|N\|\}$$

is a Banach-Lie algebra. Show that the subset

$$J =: \{(i\rho I, -i\rho I);\quad \rho\in\mathbb{R}\}$$

is a closed Lie ideal of $iHer(H_2,H_1)$.

(4) Prove that the Lie algebra $aut^0 B(F_1)$ of the Lie group $Aut^0 B(F_1)$ is isomorphic to the quotient $iHer(H_2,H_1)/J$ endowed its quotient structure and the norm

$$\|(iM,iN)+J\| = \inf_{\rho\in I}\{\|M+\rho I\|, \|N-\rho I\|\}$$

§8.- Description of $Aut^0 B(F_k)$ and $aut^0 B(F_k)$ k= 2,3,4.

We denote by $M =: \{\begin{pmatrix}\alpha_1 & \beta\\ \beta & \alpha_2\end{pmatrix};\ \alpha_j, \beta\in\mathbb{C}\}$ the J^*-algebra of all 2×2 symmetric matrices with complex entries. As usually, whenever a null entry appears in a matrix $A\in M$ we leave a blank in the corresponding place. Any element of M can be uniquely represented in the form

$$\begin{pmatrix}\alpha_1 & \beta\\ \beta & \alpha_2\end{pmatrix} = \begin{pmatrix}\alpha_1 & \cdot\\ \cdot & \alpha_2\end{pmatrix} + \beta\begin{pmatrix}\cdot & 1\\ 1 & \cdot\end{pmatrix}$$

i.e., as a sum of a diagonal matrix plus a scalar multiple of a particular element of M . First we look for the J^*-automorphisms of M that preserve diagonal matrices.

8.47. LEMMA. *Let $\Lambda: M\to M$ be a surjective linear isometry of M that preserves diagonal matrices. Then Λ is of the form*

$$\Lambda: \begin{pmatrix}\alpha_1 & \beta\\ \beta & \alpha_2\end{pmatrix} \to \begin{pmatrix}\alpha_1 & \cdot\\ \cdot & \alpha_2\end{pmatrix} + \beta Z, \quad A = \begin{pmatrix}\alpha_1 & \beta\\ \beta & \alpha_2\end{pmatrix}\in M$$

where $Z = \sigma\begin{pmatrix}\cdot & 1\\ 1 & \cdot\end{pmatrix}$ *for some* $\sigma\in\{-1,1\}$.

Proof: We must have

$$Z=\Lambda\begin{pmatrix}\cdot & 1\\ 1 & \cdot\end{pmatrix}=\Lambda\left[-\begin{pmatrix}1 & \cdot\\ \cdot & -1\end{pmatrix}^{*}\begin{pmatrix}\cdot & 1\\ 1 & \cdot\end{pmatrix}\begin{pmatrix}1 & \cdot\\ \cdot & -1\end{pmatrix}^{*}\right]=$$

$$=\left[-\Lambda\begin{pmatrix}1 & \cdot\\ \cdot & -1\end{pmatrix}\right]^{*}\Lambda\begin{pmatrix}\cdot & 1\\ 1 & \cdot\end{pmatrix}\left[\Lambda\begin{pmatrix}1 & \cdot\\ \cdot & -1\end{pmatrix}\right]^{*}=-\begin{pmatrix}1 & \cdot\\ \cdot & -1\end{pmatrix}Z\begin{pmatrix}1 & \cdot\\ \cdot & -1\end{pmatrix}$$

Then, if $Z=\begin{pmatrix}a_1 & b\\ b & a_2\end{pmatrix}$ it follows that $\begin{pmatrix}a_1 & b\\ b & a_2\end{pmatrix}=\begin{pmatrix}-a_1 & b\\ b & -a_2\end{pmatrix}$ whence

$Z=\zeta\begin{pmatrix}\cdot & 1\\ 1 & \cdot\end{pmatrix}$ for some $\zeta\in\mathbb{C}$. Clearly $|\zeta|=1$

$$1=\left\|\begin{pmatrix}\cdot & 1\\ 1 & \cdot\end{pmatrix}\right\|=\left\|\Lambda\begin{pmatrix}\cdot & 1\\ 1 & \cdot\end{pmatrix}\right\|=\|Z\|=|\zeta|$$

On the other hand, we must have (cf. lemma 8.32 and theorem 8.39)

$$\Lambda[(1\ 1)\otimes(1,1)^{*}]=\Lambda\begin{pmatrix}1 & 1\\ 1 & 1\end{pmatrix}=\Lambda\begin{pmatrix}1 & \cdot\\ \cdot & 1\end{pmatrix}+\Lambda\begin{pmatrix}\cdot & 1\\ 1 & \cdot\end{pmatrix}=\begin{pmatrix}1 & \zeta\\ \zeta & 1\end{pmatrix}=(\alpha\ \beta)\otimes(\alpha\ \beta)^{*}=$$

$$=\begin{pmatrix}\alpha^2 & \alpha\beta\\ \alpha\beta & \beta^2\end{pmatrix}$$

for suitable $\alpha,\beta\in\mathbb{C}$. But then $\alpha,\beta\in\{-1,1\}$; thus $\zeta=\alpha\beta\in\{-1,1\}$.

#

Denote by $F_2=\{X\in L(H);\ X^{\tau}=X\}$ the Cartan factor of type II corresponding to a given conjugation Q on H, where $X^{\tau}=QX^{*}Q$.

<u>8.48. THEOREM.</u> *Let* $L\in \mathrm{Aut}^{0}B(F_2)$ *be given. Then there exists a unique unitary operator* $U\in L(H)$ *such that*

$$L(A)=UAU^{\tau}\qquad A\in F_2$$

Proof: From the characterization of the minimal elements of F_2 it follows that

$$(8.30) \qquad L[e\otimes(Qe)^*] = (Te)\otimes(QTe)^* , \qquad e\in H$$

for some not necessarily linear mapping $T: H\to H$. Observe that the values of T are determined up to a constant factor belonging to $\{-1,1\}$; furthermore,

$$\|Tx\| = \|x\| , \quad T(\lambda x)\in\{-\lambda,\lambda\}Tx, \quad \lambda\in\mathbb{C}, \quad x\in H$$

Given any $u,v\in H$, we have

$$[u\otimes(Qu)^*][v\otimes(Qv)^*]^*[u\otimes(Qu)^*] = \langle u,v\rangle^2 u\otimes(Qv)^*$$

whence, by (8.30), it follows that

$$\langle Tu,Tv\rangle^2 (Tu)\otimes(QTu)^* = [(Tu)\otimes(QTu)^*][(Tv)\otimes(QTv)^*]^*[(Tu)\otimes(QTu)^*] =$$
$$= \langle u,v\rangle^2 (Tu)\otimes(QTu)^*$$

That is,

$$\langle Tu,Tv\rangle^2 = \langle u,v\rangle^2, \qquad u,v\in H$$

and, in particular, the mapping T preserves orthogonality. As a consequence we obtain that

$$(8.31) \quad L[u_1\otimes(Qu_2)^*+u_2\otimes(Qu_1)^*]\in \sum_{j,k=1}^{2} \mathbb{C}(Tu_j)\otimes(QTu_k)^* , \quad u_1,u_2\in H$$

Indeed, we have

$$Tf \perp T(\mathbb{C}u_1+\mathbb{C}u_2) \quad \text{whenever} \quad f\perp u_1,u_2$$

Observe that, if $h\perp Tu_1,Tu_2$ then, for some $f\in H$, we have

$$L[f\otimes(Qf)^*] = h\otimes(Qh)^* , \quad \text{i.e.} \quad Tf\in\{-h,h\} \quad \text{and} \quad f\perp u_1,u_2$$

(because $\langle f,u_j\rangle^2 = \langle Tf,Tu_j\rangle^2 = \langle h,Tu_j\rangle^2 = 0$ for $j= 1,2)$.

Therefore $h \perp T(\mathbb{C}u_1+\mathbb{C}u_2)$ whenever $h \perp Tu_1, Tu_2$, i.e.

$$\{Tu_1, Tu_2\}^{\perp} \subset [T(\mathbb{C}u_1+\mathbb{C}u_2)]^{\perp}$$

and so

$$\mathbb{C}Tu_1+\mathbb{C}Tu_2 \supset T(\mathbb{C}u_1+\mathbb{C}u_2) \tag{8.32}$$

Now

$$L[u_1 \otimes (Qu_2)^* + u_2 \otimes (Qu_1)^*] \in$$

$$\in \mathrm{Span}\{L[(\alpha_1 u_1+\alpha_2 u_2) \otimes Q(\alpha_1 u_1+\alpha_2 u_2)^*];\ \alpha_1, \alpha_2 \in \mathbb{C}\} \subset$$

$$\subset \mathrm{Span}\{[T(\alpha_1 u_1+\alpha_2 u_2)] \otimes [QT(\alpha_1 u_1+\alpha_2 u_2)]^*;\ \alpha_1, \alpha_2 \in \mathbb{C}\}$$

whence (8.31) follows by (8.32).

Consider now any orthonormal couple $\{e_1, e_2\}$. From (8.31) we can see the existence of a linear mapping

$$\Lambda:\ (\alpha_{jk})_{jk=0,1} \longrightarrow (\alpha'_{jk})_{jk=0,1} \qquad (\alpha_{jk}), (\alpha'_{jk}) \in M$$

of symmetric matrices such that

$$L \sum_{j,k} \alpha_{jk} e_j \otimes (Qe_k)^* = \sum_{j,k} \alpha'_{jk} (Te_j) \otimes (QTe_k)^*$$

Since for any orthonormal couple $\{h_1, h_2\}$ we have

$$\left\| \begin{pmatrix} \alpha_1 & \beta \\ \beta & \alpha_2 \end{pmatrix} \right\| = \| \alpha_1 h_1 \otimes (Qh_1)^* + \beta h_1 \otimes (Qh_2)^* + \beta h_2 \otimes (Qh_1)^* + \alpha_2 h_2 \otimes (Qh_2)^* \|$$

it follows from lemma 8.47 that

$$\Lambda \begin{pmatrix} \alpha_1 & \beta \\ \beta & \alpha_2 \end{pmatrix} = \begin{pmatrix} \alpha_1 & \sigma\beta \\ \sigma\beta & \alpha_2 \end{pmatrix}, \qquad \begin{pmatrix} \alpha_1 & \beta \\ \beta & \alpha_2 \end{pmatrix} \in M$$

holds for some $\sigma \in \{-1,1\}$. Thus, the isometry defined by

$$U_{e_1 e_2} : \zeta_1 e_1 + \zeta_2 e_2 \longrightarrow \zeta_1 Te_1 + \zeta_2 Te_2 , \qquad \zeta_1 \zeta_2 \in \mathbb{C}$$

satisfies

$$L\{(\zeta_1 e_1 + \zeta_2 e_2) \otimes [Q(\zeta_1 e_1 + \zeta_2 e_2)]^*\} = [U_{e_1 e_2}(\zeta_1 e_1 + \zeta_2 e_2)] \otimes$$
$$\otimes [QU_{e_1 e_2}(\zeta_1 e_1 + \zeta_2 e_2)]^*$$

Finally, let us fix a unit vector $e_0 \in H$. Observe that $U_{e_0, \lambda f} = U_{e_0, f}$ whenever $f \perp e_0$, $\| f \| = |\lambda| = 1$. Indeed, we have dom $U_{e_0, \lambda f}$ = dom $U_{e_0, f}$; moreover, as T is pointwise determined up to a constant ± 1, $U_{e_0, \lambda f}$ is unambiguously determined by its linearity and the fact $U_{e_0, \lambda f}(e_0) = Te_0$ for $\lambda \in \mathbb{C}$. Hence, we can define a mapping T' on H by means of

$$T'(\alpha e_0 + \beta f_0) =: U_{e_0 f_0}(\alpha e_0 + \beta f_0), \qquad f_0 \perp e_0, \ \| f_0 \| = 1, \quad \alpha, \beta \in \mathbb{C}$$

Also, if the subscript x denotes the value at $x \in H$, T' satisfies

$$(8.33) \qquad L[e_0 \otimes (Qf)^* + f \otimes (Qe_0)^*]_x = T'e_0 \otimes (QT'f)^*_x + (T'f) \otimes (QT'e_0)^*_x$$

for all $f, x \in H$. The left han side of (8.33) is a linear mapping of the variable f. It follows that the mapping $f \to T'f$ is linear, too. Indeed, given $f, g \in H$, we can write $f = \alpha e_0 + f_0$, $g = \beta e_0 + g_0$ for some $\alpha, \beta \in \mathbb{C}$, and $f_0, g_0 \in H$ with $f_0, g_0 \perp e_0$. Now, the orthogonal projection of the right hand side of (8.33) onto the subspace $\{e_0\}^{\perp}$ is equal to $(T'f_0) \otimes (QT'e_0)^*_x$ for any $x \in H$, whence

$$(T'f_0) \otimes (QT'e_0)_x + (T'g_0) \otimes (QT'e_0)^*_x = [T'(f_0 + g_0)] \otimes (QT'e_0)^*_x . \qquad x \in H$$

This implies $T'f_0+T'g_0 = T'(f_0+g_0)$. On the other hand,

$$T'f = \alpha T'e_0+T'f_0 \qquad\qquad T'g = \beta T'e_0+T'g_0$$

Since $T'f\in\{-Tf,Tf\}$ for all $f\in H$, it follows that T' is a linear isometry of H such that (8.30) holds with T' instead of T. This completes the proof, because then

$$L[e\otimes(Qe)^*] = T'\circ[e\otimes(Qe)^*]\circ T'^{\tau}, \qquad e\in H$$

holds and, by theorem 8.41, the latter entails

$$L(A) = T'AT'^{\tau} \qquad\qquad A\in F_2$$

#

<u>8.49. EXERCISES</u>. (1) If $\{h_1,h_2,\ldots,h_n\}$ is an orthonormal system in H and $(\alpha_{jk})_{j,k=1,\ldots,n}$ is any complex symmetric matrix, then

$$\|(\alpha_{jk})_{jk}\| = \|\sum_{j,k=1}^{n} \alpha_{jk}h_j\otimes(Qh_k)^*\|$$

(2) Show that the Lie group $Aut^0B(F_2)$ is connected and that its Lie algebra is given by

$$aut^0B(F_2) = \{f_M\colon X \to i(MX+XM^{\tau});\quad M\in Her(H)\}$$

Hint: Look for continuous one-parameter groups

$$s\to L_s \text{ of the form } \quad L_s(A) = U_sAU_s^{\tau}, \quad A\in F_2, \quad s\in\mathbb{R}$$

(cf. example 6.43).

<u>8.50. REMARK</u>. *It would be interesting to find representation formulas for the elements of the groups* $Aut^0B(F_3)$ *and* $Aut^0B(F_4)$. *It seems that the method presented for* $Aut^0B(F_2)$ *can be adapted to the case of* $Aut^0(F_3)$. *However, the*

study of Cartan factors of type IV, and in particular that of $\mathrm{Aut}^0 B(F_4)$, *requires a different approach (cf. |24| and |74| respectively).*

CHAPTER 9

BOUNDED SYMMETRIC DOMAINS

§1.- Historical sketch.

9.1. DEFINITION. *We say that a bounded domain* D *in a complex Banach space* E *is "symmetric at a point"* $a\in D$ *if there exists an automorphism* $S\in \mathrm{Aut}(D)$ *such that* $S^2 = id_D$ *and* a *is an isolated fixed point for* S.

We say that D *is "symmetric" if it is symmetric at every point* $a\in D$.

Throughout the whole chapter, E and D denote respectively a fixed complex Banach space and a bounded *simply connected* symmetric domain D in E.

Symmetric domains in $\mathbb{C}^n$ where introduced by E. Cartan in 1935 who pointed out the very deep consequences of this definition by classifying completely all possible finite dimensional symmetric domains. His proofs were based on the complete classification of finite dimensional semisimple Lie algebras (given also by himself previously), a tool whose use seems rather hopeless in infinite dimensions. Since we are primarily interested in infinite dimensions, we state Cartan's theorem without proof.

Let $E_1,..,E_n$ be finite dimensional complex Banach spaces of dimensions $r_1,..,r_n$ and write $\bigoplus_{k=1}^{n} E_k$ for the space

$$\bigoplus_{k=1}^{n} E_k =: \{(x_1,..,x_n);\ x_k\in E_k \quad \forall k= 1,..,n\}$$

endowed with the norm

$$\| (x_1,..,x_n) \| =: \max\{ \| x_1 \|_{E_1},..., \| x_n \|_{E_n} \}$$

Then we have:

THEOREM (Cartan's classification theorem). *If* dim $E<\infty$, *then there are Banach spaces* $E_1,..,E_n$ *such that* D *is biholomorphically equivalent to the open unit ball* $B(\bigoplus_{k=1}^{n} E_k)$ *of* $\bigoplus_{k=1}^{n} E_k$, *where each of the* E_k *is equal to some of the spaces described below (ordered in 6 fundamental types):*

TYPE I: $L(\mathbb{C}^p,\mathbb{C}^q)$ *whit* $p, q \in \mathbb{N}$

TYPE II: $\{A \in L(\mathbb{C}^p,\mathbb{C}^q);\ A^t = A\}$ *where* $p,q \in \mathbb{N}$ *and* A^t *denotes the transposed of* A.

TYPE III: $\{A \in L(\mathbb{C}^p,\mathbb{C}^q);\ A^t = -A\}$ *with* $p,q \in \mathbb{N}$.

In these cases, $\mathbb{C}^p$, $\mathbb{C}^q$ *and* $L(\mathbb{C}^p,\mathbb{C}^q)$ *have, respectively, their euclidean norm and the operator norm*

TYPE IV: *The space* $\mathbb{C}^p$ *with the norm*

$$\| (\zeta_1,...,\zeta_p) \|^2 =: \sum_{j=1}^{p} |\zeta_j|^2 + \left(\left(\sum_{j=1}^{p} |\zeta_j|^2\right)^2 - \left| \sum_{j=1}^{p} \zeta_j^2 \right|^2 \right)^{\frac{1}{2}}$$

TYPE V: *A particular Banach space of dimension 16.*

TYPE VI: *A particular Banach space of dimension 27.*

From chapter 7 we know that this representation of D is unique up to an isometric linear isomorphism. The domain $B(\bigoplus_{k=1}^{n} E_k)$ is called the *Harish-Chandra realization* of D.

If we drop the condition dim $E<\infty$, the situation seems to become essentially more complicated. So far, the strongest analogous general result is a very recent theorem of W. Kaup (1983) |35|:

THEOREM. *Let* D *be a bounded symmetric domain of* E. *Then* D *is biholomorphically equivalent to the unit open ball of some Banach space which is uniquely determined up to isometric linear isomorphisms.*

In this chapter, we prove only a slyghtly weaker statement due to J.P. Vigué (1976):

THEOREM. *Let* D *be a bounded symmetric domain of* E. *Then, there exists a balanced domain* D' *in* E *such that* D *and* D' *are biholomorphically equivalent.*

Since biholomorphically equivalent balanced domains are linearly equivalent, Kaup's theorem shows the convexity of D'. However, from the way of Vigué's construction this fact cannot be discovered.

Finally, we remark that several ideas of Kaup's method go back to a modern elementary Jordan theoretic approach of the school of M. Koecher (1969) to finite dimensional symmetric domains which we recommend to the interested readers, (cf., |42| and |43|).

§2.- Elementary properties of symmetric domains.

The remaining paragraphs contain the proof of Vigué's theorem divided into steps that might have some interest in themselves.

9.2. LEMMA. *Given* $a \in D$ *and* $S \in \mathrm{Aut}D$ *such that* $S_a = a$ *and* $S^2 = id_D$, *there exist a neighbourhood* U *of* a *and a biholomorphic map* $f\colon U \to E$ *such that* $f(a) = 0$ *and* $f_{\#}S$ *is linear.*

Proof: We may assume $a = 0$. Write $L =: S_0^{(1}$ and remark that $L^2 = id$. For $h \in \mathrm{Hol}(D,E)$, we define $\psi(h) =: L \circ h \circ S$. Then the operator ψ is linear and satisfies $\psi^2 = id$. Hence, the mapping $g =: \frac{1}{2}(S + \psi(S))$ is a fixed point of ψ, i.e. $g = L \circ g \circ S$. Furthermore $g(0) = 0$ and $g_0^{(1} = \frac{1}{2}(L+L) = L$. Since $g_0^{(1}$ is invertible, there is a neighbourhood V of 0 where g is biholomorphic. Put $U = V \cap S(V)$; then U is a neighbourhood of 0 such that $S(U) = U$. By setting $g =: g_{|U}$ we have $f(0) = 0$ and $f = L \circ f \circ S$, i.e. $F_{\#}S = f \circ S \circ f^{-1} = L_{|f(U)}$.

#

9.3. PROPOSITION. *For every* $a\in D$, *any symmetry* S *of* D *at* a *satisfies* $S(a) = a$ *and* $S_a^{(1} = -\mathrm{id}$. *In particular,* S *unique.*

Proof: We assume $a = 0$. Let us denote by S any symmetry of D at a. By lemma 9.2 we have a local coordinate map f at 0 such that $f(0) = 0$ and $L =: S_0^{(1}$ is a linear continuation of $f_{\#}S$. We show that $L = -\mathrm{id}$. Assume we had $Lx \neq -x$ for some $x\in E$, $x\neq 0$. Then, for sufficiently small values of t (say $|t|<\delta$), the points $x_t =: t(x+Lx)$ lie in the neighbourhood where f is defined. Moroever, as $L^2 = \mathrm{id}$, we have $L(x_t) = x_t$ for all $t\in(-\delta,+\delta)$. But then $f^{-1}(x_t)$ is a fixed point for S. Since $\lim_{t\to 0} f^{-1}(x_t) = f^{-1}(0)=0$ and $x_t\neq 0$ for $t\neq 0$, S cannot have an isolated fixed point. By Cartan's uniqueness theorem, S is uniquely dermined. #

9.4. DEFINITION. *We shall designate the symmetry of* D *at* a *by* S_a.

9.5. PROPOSITION. *The mapping* $D\to \mathrm{Aut}D$ *given by* $a\to S_a$ *is* T*-continuous.*

Proof: Given $a\in D$, we show that

$$(9.1) \qquad T\lim_{c\to 0} S_{a+c}\, S_a = \mathrm{id}_D$$

Then, multiplying on the right by $S_a^{-1} = S_a$ and applying theorem 2.2 we obtain $T\lim_{c\to 0} S_{a+c} = S_a$.

By theorem 2.8, in order to prove (9.1) it suffices to show that

$$\lim_{c\to 0} (S_{a+c}\, S_a)_a^{(k} = \mathrm{id}_a^{(k} \quad \text{for } k= 0,1$$

Now we have

$$(S_{a+c}\, S_a)_a^{(0} = S_{a+c}(S_a(a)) = S_{a+c}(a) =$$

$$= S_{a+c}(a+c) + [S_{a+c}(a) - S_{a+c}(a+c)] = a+c+[S_{a+c}(a)-S_{a+c}(a+c)] =$$

$$= a+c-\int_0^1 (S_{a+c})^{(1}_{a+tc}\, c\, dt$$

Take a ball $B\subset\subset D$ centered at a and put $\delta =: \mathrm{dist}(B,\partial D)$. From the Cauchy estimates we have

$$\|(S_{a+c})^{(1}_{a+tc}\| \leqslant \frac{1}{\delta}\|S_{a+c}\|_D$$

Since $S_{a+c}(D)\subset D$ and D is bounded, we get

$$\|(S_{a+c}\, S_a)^{(1}_a - a\| \leqslant \|c\|\,(1+\frac{1}{\delta}M)$$

for some M independent of c. Therefore, $\lim_{c\to 0}(S_{a+c}\, S_a)^{(0}_a = a = \mathrm{id}^{(0}_a$. On the other hand,

$$(S_{a+c}\, S_a)^{(1}_a - \mathrm{id} = (S_{a+c})^{(1}_{S_a(a)}(S_a)^{(1}_a - \mathrm{id} = (S_{a+c})^{(1}_a + (S_{a+c})^{(1}_{a+c} =$$
$$= \int_0^1 (S_{a+c})^{(2}_{a+tc}\, c\, dt$$

From the Cauchy estimates,

$$\|(S_{a+c})^{(2}_{a+tc}\| \leqslant (\frac{e}{\delta})^2\|S_{a+c}\|_D = (\frac{e}{\delta})^2 M$$

so that

$$\|(S_{a+c}\, S_a)^{(1}_a - \mathrm{id}^{(1}_a\| \leqslant \|c\|\,(\frac{e}{\delta})^2 M \to 0$$

#

Henceforth we assume that $0\in D$ and write $S =: S_0$ for the symmetry of D at 0.

<u>9.6. PROPOSITION</u>. *Let* $a\in D$ *be given. Then, for every* $c\in E$, *there exists a unique* $A_c \in \mathrm{aut}D$ *such that*

$$(A_c)_a^{(0} = c \qquad (A_c)_a^{(1} = \frac{1}{2} S_a^{(2}(c,.).$$

Namely such a vector field is given by

$$A_c = T \lim_{t\to 0} \frac{1}{2t} (S_{a+ct} \; S-id_D)$$

Proof: We may assume a= 0. In order to utilize theorem 5.8, we study the convergence of $\frac{1}{2t}(S_{tc} \; S\text{-id})_0^{(k}$, k= 0,1, for $t\to 0$. We shall make use of the following Taylor type formulas for holomorphic maps F:

$$(9.2) \quad F(b+h) = F(b)+F(b+h)-F(b) = F_b^{(0} + \int_0^1 F_{b+\tau h}^{(1} h d\tau =$$

$$= F_b^{(0} + F_b^{(1} h + \int_0^1 (F_{b+\tau h}^{(1} - F_b^{(1})h \, d\tau =$$

$$= F_b^{(0} + F_b^{(1} h + \int_0^1\int_0^1 F_{b+\xi\tau h}^{(2}(h,\tau h)\, d\xi d\tau$$

Let us take any ball $B \subset\subset D$ centered at 0 and put $\delta =: \mathrm{dist}(B,\partial D)$. For sufficiently small values of t, we have $ta \in B \subset D$ and applying (9.2) to the function $F =: S_{tc}$ and the points $b =: tc$, $h =: -tc$, we obtain

$$S_{tc}(0) = S_{tc}(tc) + (S_{tc})_{tc}^{(1}(-tc) + \int_0^1\int_0^1 (S_{tc})_{t(1-\xi\tau)c}^{(2}(tc,-tc)\, d\xi d\tau =$$

$$= 2tc - t^2 \int_0^1\int_0^1 (S_{tc})_{t(1-\xi\tau)c}^{(2}(c,c)\, d\xi d\tau$$

Moreover, as $t(1-\xi\tau)c \in B$ for all $\tau,\xi \in [0,1]$, from the Cauchy estimates we obtain

$$\| (S_{tc})_{t(1-\xi\tau)c}^{(2} \| \leqslant \left(\frac{e}{\delta}\right)^2 \| S_{tc} \|_D = \left(\frac{e}{\delta}\right)^2 M$$

where M is independent of t. Therefore,

$$\left\| \frac{1}{2t} (S_{tc} \; S-id_D)_0^{(0} - c \right\| = \left\| \frac{1}{2t} S_{tc}(0) - c \right\| \leq \frac{|t|}{2} \|c\|^2 \left(\frac{e}{2}\right)^2 M \to 0$$

so that $\lim_{t\to 0} \frac{1}{2t} (S_{tc} \; S-id_D)_0^{(0} = c.$

On the other hand, applying (9.2) to the operator-valued function $f\colon x \to (S_{tc})_x^{(1}$ and the points $b=: tc$, $h=-tc$, we have

$$(S_{tc})_0^{(1} = (S_{tc})_{tc}^{(1} + (S_{tc})_{tc}^{(2}(-tc) + \int_0^1\int_0^1 (S_{tc})_{t(1-\xi\tau)c}^{(3}(-tc,-\tau tc)\,d\xi d\tau =$$

$$= -id - t(S_{tc})_{tc}^{(2}(c) + t^2\int_0^1\int_0^1 (S_{tc})_{t(1-\xi\tau)c}^{(3}(c,c)\,d\tau d\xi$$

Again by the Cauchy estimates,

$$\| (S_{tc})_{t(1-\xi\tau)c}^{(3} \| \leq \left(\frac{e}{\delta}\right)^3 \| S_{tc} \|_D = \left(\frac{e}{\delta}\right)^3 M$$

But now

$$(S_{tc} \; S-id_D)_0^{(1} = (S_{tc})_{S_0(0)}^{(1} S_0^{(1} - id = -(S_{tc})_0^{(1} - id$$

so that

$$\left\| \frac{1}{2t} (S_{tc} \; S-id_D)_0^{(1} - \frac{1}{2} (S_{tc})_{tc}^{(2} c \right\| \leq \frac{|t|}{2} \left(\frac{e}{\delta}\right)^3 M \to 0$$

Moreover, by proposition 9.5 we have $\mathcal{T}\lim_{t\to 0} S_{tc} = S$ which implies $\lim_{t\to 0} (S_{tc})_{tc}^{(2} = S_0^{(2}$.

Thus, theorem 5.8 entails that

$$A_c =: \mathcal{T}\lim_{t\to 0} \frac{1}{2t} (S_{tc} \; S-id_D)$$

defines a vector field $A_c \in autD$ which obviously satisfies the requirements of the statement.

#

9.7. DEFINITION. *Henceforth, for* c∈E, A_c *will denote the vector field of* autD *uniquely determined by the conditions*

(9.3) $(A_c)_0^{(0} = 0$ $(A_c)_0^{(1} = \frac{1}{2} S_0^{(2}(c, \cdot)$

It is easy to show that $\psi : E \to$ autD given by $\psi =: c \to A_c$ is a continuous real linear mapping.

9.8. PROPOSITION. *Let* D *be a bounded symmetric domain. Then* D *is homogeneous under the action of the subgroup* $Aut_0 D$.

Proof: By proposition 7.17, it suffices to show that the orbit $(Aut_0 D)0$ of the origin by the subgroup $Aut_0 D$ is a neighbourhood of 0. By lemma 6.45 and remark 6.2, we can find a neighbourhood M of the origin in autD such that the mapping $A \in M \to expA \in AutD$ is real analytic, and $expM \subset Aut_0 D$. Since $\psi : c \to A_c$ is real analytic, we can find a neighbourhood U of 0 in E such that $\psi(U) \subset M$. Moreover, by theorem 6.57, $f \in AutD \to f(0) \in D$ is real analytic. Therefore, the composite

$$g =: c \in U \to (expA_c)0 \in D$$

is real analytic in U and we have

$$g(U) = exp\psi(U)0 \subset M(0) \subset (Aut_0)0$$

Besides, $g(0) = 0$ and

$$g_0^{(1}c = \frac{d}{dt} \Big|_0 (expA_{tc})0 = \frac{d}{dt} \Big|_0 (exptA_c)0 = A_c(0) = c$$

for all c∈E, so that $g_0^{(1} = $ id. Then, by the inverse mapping theorem, g(U) is a neighbourhood of the origin.

                                                                    #

§3.- The canonical decomposition of autD.

By investigating the effect of the adjoint $S_{\#}$ of the symmetry at 0, we obtain a good picture of autD. We already know that autD is a real Banach space and that $S_{\#}$: autD → autD is a continuous linear operator. Moreover, as $S^2 = id_D$, $S_{\#}$ is a projector, so that it furnishes a decomposition of autD into the direct topological sum

$$\text{autD} = \mathcal{L} \oplus \mathcal{Q}$$

of the eigensubspaces corresponding to the eigenvalues -1 and +1 of $S_{\#}$,

$$(9.4) \qquad \begin{aligned} \mathcal{L} &=: \{A \in \text{autD};\ S_{\#}A = -A\} \\ \mathcal{Q} &=: \{A \in \text{autD};\ S_{\#}A = A\} \end{aligned}$$

9.9. PROPOSITION. *The subspaces $\mathcal{L}$ and $\mathcal{Q}$ are, respectively, the image and the kernel of the mappings*

$$\psi: E \to \text{autD} \qquad \varphi: \text{autD} \to E$$
$$c \to A_c \qquad\qquad A \to A(0)$$

Proof: Let A∈autD be given. By definition of $S_{\#}$, we have

$$(S_{\#}A)x = S^{(1}[S^{-1}(x)]A[S^{-1}(x)]$$

for all x∈D. By taking the first derivative, we get

$$(S_{\#}A)^{(1}x = S^{(1}[S^{-1}(x)]A^{(1}[S^{-1}(x)]S^{(1}(s) + S^{(2}[S(x)]S^{(1}(x)A[S(x)]$$

Thus, especializing these relations for x= 0 and taking into account that $S(0) = 0$, $S^{(1}(0) = -id$, we obtain

$$(9.5) \qquad (S_{\#}A)_0^{(0} = -A_0^{(0} \qquad (S_{\#}A)_0^{(1} = A_0^{(1} - S_0^{(2}(A(0), \cdot)$$

for all A∈autD.

Now we prove $L = \mathrm{Im}\psi$. Let $A \in \mathrm{aut}D$ be such that $S_{\#}A = -A$ and define $c =: A_0^{(0}$. Then we have $(S_{\#}A)_0^{(1} = -A_0^{(1}$ and, by (9.5), $(S_{\#}A)_0^{(1} = A_0^{(1} - S_0^{(2}(c,.)$ so that $A_0^{(1} = \frac{1}{2} S_0^2(c,.)$. Hence by definition 9.7, we get $A = A_c$.

Conversely, let us assume that $A = A_c$ for some $c \in E$, so that $A_0^{(0} = c$ and $A_0^{(1} = \frac{1}{2} S_0^{(2}(c,.)$. By (9.5) the vector field $S_{\#}A$ satisfies

$$(S_{\#}A)_0^{(0} = -c\ , \qquad (S_{\#}A)_0^{(1} = -\frac{1}{2} S_0^{(2}(c,.)$$

By Cartan's uniqueness theorem we have $S_{\#}A = -A$.

The proof of $Q = \ker\varphi$ is quite similar.

#

9.10. COROLLARY. *We have*

$$[L,L] \subset L \qquad [L,Q] \subset Q \qquad [Q,Q] \subset Q$$

In particular, L *is a closed Lie subalgebra of* autD. *Moreover,*

$$[L,A_c] = A_{L_0^{(1}c}$$

for all $A_c \in Q$ *and* $L \in L$.

Proof: Since $S\#$ is a Lie algebra automorphism, we have

$$S_{\#}[A_u,A_v] = [S_{\#}A_u,S_{\#}A_v] = [-A_u,-A_v] = [A_u,A_v]$$

$$S_{\#}[A_u,L] = [S_{\#}A_n,S_{\#}L] = [-A_u,L] = -[A_u,L]$$

$$S_{\#}[L_1,L_2] = [S_{\#}L_1,S_{\#}L_2] = [L_1,L_2].$$

Furthermore,

$$[L,A_c]0 = L_0^{(1}A_c0 - (A_c)_0^{(1}L0 = L_0^{(1}c.$$

#

§4.- The complexified Lie algebra of autD.

For later use, we introduce the complexified

$$\mathbb{C}\mathrm{autD}=: (\mathrm{autD})\oplus i(\mathrm{autD})$$

of the Lie algebra autD. Since autD is purely real, i.e. $(\mathrm{autD})\cap i(\mathrm{autD})=\{0\}$, this represents $\mathbb{C}$autD as a direct topological sum. Thus, though the vector fields $A\in\mathbb{C}\mathrm{autD}$ are no longer complete in D (and, in particular, they do not satisfy Cartan's uniqueness theorem), they have a unique representation of the form $X= A_1+iA_2$ with $A_1, A_2\in\mathrm{autD}$.

9.11. DEFINITION. *For* $c\in E$, *we define*

$$C_c=: \frac{1}{2}(A_c-iA_{ic}) \qquad Q_c=: \frac{1}{2}(A_c+iA_{ic})$$

Therefore, we have

$$(9.6) \qquad C_c(0)=c \qquad Q_c(0)=0 \qquad A_c= C_c+Q_c$$

for all $c\in E$. Moreover, the mappings $E\to\mathbb{C}\mathrm{autD}$ given by $c\to C_c$ and $c\to Q_c$ are, respectively, complex linear and complex conjugate linear. Both of them are continuous.

9.12. PROPOSITION. *We have* $[A_c,A_{ic}]\neq 0$ *for all* $c\in E$, $c\neq 0$.

Proof: Suppose we had $[A_c,A_{ic}]=0$ for some $c\in E$, $c\neq 0$. Fix any ball $B\subset\subset D$ centered at 0 and assume that $\mathbb{C}$autD has been endowed with the norm $\|\cdot\|_B$. Then, there exists $\varepsilon>0$ such that

$$\exp X|_B=\sum_{k=0}^{\infty}\frac{1}{k!}\hat{X}^k\,\mathrm{id}|_B$$

is well defined on B whenever $X\in\mathbb{C}\mathrm{autD}$ and $\|X\|<\varepsilon$. Since the mappings $\lambda\to A_{\lambda c}$ and $\lambda\to C_{\lambda c}$ are continuous, there is some $\delta>0$ such that we have

$$\|A_{\lambda c}\| < \varepsilon/2 \quad , \quad \|C_{\lambda c}\| < \varepsilon/2$$

for all $\lambda \in \mathbb{C}$, $|\lambda| \leq \delta$. Therefore, we can define

$$f(\lambda) =: (\exp A_{\lambda c})^{-1}(\exp C_{\lambda c})0, \quad |\lambda| \leq \delta$$

From the assumption $[A_c A_{ic}] = 0$, we easily derive $[A_{\lambda c}, A_{\lambda c}] = 0$ for all $\lambda \in \mathbb{C}$; thus by corollary 4.24, we have

$$F(\lambda) = (\exp A_{\lambda c})^{-1}(\exp C_{\lambda c})0 = \exp(-A_{\lambda c} + C_{\lambda c})0 \qquad |\lambda| \leq \delta$$

But obviously $(-A_{\lambda c} + C_{\lambda c})0 = -A_{\lambda c}(0) + C_{\lambda c}(0) = 0$ by (9.6) so that $\exp(-A_{\lambda c} + A_{\lambda c})0 = 0$ for all $|\lambda| \leq \delta$, i.e.

$$(\exp A_{\lambda c})0 = (\exp C_{\lambda c})0 \tag{9.7}$$

Moreover, since $\lambda \to C_{\lambda c}$ is complex linear and continuous, we have

$$(\exp C_{\lambda c})0 = \sum_{k=0}^{\infty} \frac{1}{k!} (C_{\lambda c}^k \mathrm{id}_B)0 = \sum_{k=0}^{\infty} \frac{\lambda^k}{k!} (C_c^k \mathrm{id}_B)0$$

and therefore the mapping $g: \lambda \to (\exp C_{\lambda c})0$ is holomorphic in $|\lambda| < \delta$. We already know that $h: \lambda \to (\exp A_{\lambda c})0$ is real analytic on the wohle $\mathbb{C}$. By (9.7), g and h coincide in an open subset of $\mathbb{C}$; thus h is an entire mapping. But $A_{\lambda c}$ is a complete vector field in D, so that $h(\lambda) = (\exp A_{\lambda c})0 \in D$ for all $\lambda \in \mathbb{C}$. By Liouville's theorem we have

$$h(\lambda) = (\exp A_{\lambda c})0 = (\exp A_0)0 = 0$$

for all $\lambda \in \mathbb{C}$. But this is a contradiction since

$$\frac{d}{d\lambda}\Big|_0 (\exp A_{\lambda c})0 = A_c(0) = c \neq 0$$

#

<u>9.13. PROPOSITION</u>. *Let* D *be a bounded symmetric domain and assume that* $A \in \mathrm{aut}D$ *satisfies* $[A,X] = 0$ *for all* $X \in \mathrm{aut}D$. *Then*

we have A= 0.

Proof: Let $A \in \text{aut}D$ be such that $[A,X]=0$ for all $X \in \text{aut}D$ and assume that we had $A \neq 0$. Then, in the decomposition $A = A_c + L$ with $A_c \in \mathcal{Q}$, $L \in \mathcal{L}$, we have two possibilites: $L \neq 0$, $L=0$.

Suppose $L \neq 0$. As $L \in \mathcal{L}$ entails $L0=0$, by Cartan's uniqueness theorem, there must exist some $c' \in E$ such that $L_0^{(1}c' \neq 0$. Then, for $X =: A_{c'} \in \text{aut}D$, we have

$$[A,X]0 = [A_c A_{c'}]0 + [L, A_{c'}]0$$

By corollary 9.10, we have $[A_c, A_{c'}] \in \mathcal{L}$ so that $[A_c, A_{c'}]0 = 0$ and $[L, A_{c'}]0 = L_0^{(1}c' \neq 0$. Therefore $[A,X] \neq 0$ which is contradictory.

Suppose $L=0$. Then $A_c \neq 0$ and $c \neq 0$. But then, by proposition 9.12, the vector field $X =: A_{ic} \in \text{aut}D$ satisfies $[A,X] = [A_c, A_{ic}] \neq 0$ which is contradictory.

#

§5.- The local representation of autD.

9.14. LEMMA. *The mapping* $c \to (\exp C_c)$ *is biholomorphic in a neighbourhood of* 0.

Proof: It is easy to see that the mapping $c \to (\exp C_c)0$ is holomorphic in a neighbourhood of the origin. On the other hand

$$\frac{d}{dt}\Big|_0 (\exp C_{tc})0 = (\hat{C}_c \text{id}_B)0 = C_c(0) = c$$

so that its Fréchet derivative at 0 is invertible. The result follows by the inverse mapping theorem.

#

9.15. DEFINITION. *Henceforth,* U *denotes a neighbourhood of* 0 *such that* $c \to (\exp C_c)0$ *admits a holomorphic inverse on* U. *Moreover, we set*

$$V=: \{(\exp C_c)0;\ c\in U\}$$

We introduce the mapping J: V → U *by means of*

$$J: (\exp C_c)0 \to c$$

Thus, we have $U= J(V)$ and $J(\operatorname{Exp} C_c)0= c$ for all $c\in U$.

As previously, we write $\mathcal{L}=: \{L\in \operatorname{aut} D;\ L0= 0\}$.

<u>9.16. PROPOSITION</u>. *We have* $(J_\# L)x= L_0^{(1}x$ *for all* $x\in U$ *and* $L\in\mathcal{L}$.

Proof: Let $L\in\mathcal{L}$ be given and set $G^t=: \exp tL$ for $t\in\mathbb{R}$. We have $T\lim_{t\to 0} G^t= \mathrm{id}_D$; therefore, we can find a number $\delta>0$ and a neighbourhood $W_1\subset V$ of the origin such that $G^t(W_1)\subset V$ for all t, $|t|<\delta$. Let us set $W_2=: J(W_1)$, so that JG^tJ^{-1} is defined on W_2 for $|t|<\delta$. Now, let $x\in W_2$ be given; from $G^{-t}0= 0$ we derive

$$(J_\# G^t)x= (JG^tJ^{-1})x= JG^t(\exp C_x)0= JG^t(\exp C_x)G^{-t}0= J[G^t_\#(\exp C_x)]0=$$

$$= J[\exp(G^t_\# C_x)]0.$$

By theorem 4.28 we have

$$G^t_\# C_x= (\exp t\, L)_\# C_x= \sum_{n=0}^{\infty} \frac{t^n}{n!} L^n_\# C_x$$

where $L_\# C_y= [L,C_y]= C_{L_0^{(1}y}$ for $y\in E$, i.e.

$$L^n_\# C_x= C_{(L_0^{(1})^n x}$$

for $n\in\mathbb{N}$. Thus,

$$(J_\# G^t)x= J[\exp C_{\exp(tL_0^{(1})x}]0= \exp t\, L_0^{(1}x$$

whence $(J_\# L)x= L_0^{(1}x$. The result follows by the identity principle. #

<u>9.17. LEMMA</u>. *We have* $[C_x,C_y]=0$, $[Q_x,Q_y]=0$ *for all* $x,y\in E$.

Proof: For any $u,v\in E$, consider

$$N_{u,v}=:\ [A_u,A_v]-[A_{iu},A_{iv}]=\ [C_u+Q_u,\ C_v+Q_v]-[iC_u-iQ_v,\ iC_v-iQ_v]=$$

$$=2[C_u,C_v]+2[Q_u,Q_v]$$

By corollary 9.10 we have $N_{u,v}\in L$, and therefore

$$J_\# N_{u,v}=2[C_u,C_v]_0^{(1}$$

since

$$[Q_u,Q_v]_0^{(1}=(Q_u^{(1}Q_v-Q_v^{(1}Q_u)_0^{(1}=$$

$$=2(Q_u)_0^{(2}((Q_v)_0^{(0},\cdot)+(Q_u)_0^{(1}(Q_v)_0^{(1}-2(Q_v)_0^{(2}((Q_u)_0^{(0},\cdot)-(Q_v)_0^{(1}(Q_u)_0^{(1}$$

and

$$(Q_c)_0^{(s}=\frac{1}{2}(A_c+iA_{ic})_0^{(s}=\frac{1}{2}(A_c)_0^{(s}+\frac{i}{2}(A_{ic})_0^{s}=0$$

if $s=0,1$ and $c\in E$.

It follows that

$$J_\# N_{ix,y}=2[C_{ix},C_y]_0^{(1}=2i[C_x,C_y]_0^{(1}=iJ_\# N_{x,y}$$

i.e.

$$N_{ix,y}=iN_{x,y}\in L\cap(iL)\subset(\mathrm{aut}D)\cap(i\ \mathrm{aut}D)=\{0\}$$

by theorem 4.25. Thus

$$0=N_{ix,y}=i[C_x,C_y]-i[Q_x,Q_y]=N_{x,y}=[C_x,C_y]+[Q_x,Q_y]$$

whence $0=[C_x,C_y]=[Q_x,Q_y]$.

9.18. PROPOSITION. *We have*

(a) $(J_{\#}C_c)x = c$ *for all* $c\in E$ *and* $x\in U$

(b) *For all* $c\in E$, *the vector field* $J_{\#}Q_c$ *is a continuous homogeneous polynomial of second degree.*

Proof: (a) Let $c\in E$ and $x\in U$ be given. Then we have

$$x = J(\exp C_x)0 = J(\exp C_x)J^{-1}0 = (J_{\#}\exp C_x)0 = \exp(J_{\#}C_x)0$$

Whence, by lemma 4.23 we derive

$$(J_{\#}C_c)x = \frac{d}{dt}\Big|_0 [\exp(tJ_{\#}C_c)]x = \frac{d}{dt}\Big|_0 (J_{\#}\exp tC_c)(J_{\#}\exp C_c)0 =$$

$$= \frac{d}{dt}\Big|_0 J_{\#}(\exp C_{tc}\exp C_x)0 = \frac{d}{dt}\Big|_0 (J_{\#}\exp C_{x+tc})0 = \frac{d}{dt}\Big|_0 (x+tc) = c$$

so that $J_{\#}C_c$ is a constant vector field of value c.

(b) Let $c\in E$, $x\in E$ and $y\in U$ be given. By the previous step we have $x = (J_{\#}C_x)y$; therefore

$$(J_{\#}Q_c)_y^{(1}x = (J_{\#}Q_c)_y^{(1}(J_{\#}C_x)y = [J_{\#}Q_c, J_{\#}C_x]y = J_{\#}[Q_c, C_x]y$$

Moreover, we have $[Q_c, C_x]\in\mathbb{C}\mathcal{L}$ so that

$$[Q_c, C_x] =: L_{c,x}$$

is a vector field to which proposition 9.16 applies. Thus, we have

$$(9.8) \qquad (J_{\#}L_{c,x})y = (L_{c,x})_0^{(1}$$

Now, for x in a neighbourhood of the origin, the segment $[0,x]$ lies in U and we can define $\phi: [0,1] \to E$ by means of $\phi(t) = (J_{\#}Q_c)tx$. It is easy to check that $\phi(0) = 0$, so that by (9.8) we have

$$(J_{\#}Q\ )x=\phi(1)-\phi(0)=\int_0^1 \frac{d}{dt}\phi(t)dt=\int_0^1 (J_{\#}Q_c)\Big(^1_{tx}xdt=$$

$$=\int_0^1 (L_{c,x})\Big(^1_0 txdt=\frac{1}{2}(L_{c,x})\Big(^1_0 x=\frac{1}{2}\left[Q_c,C_x\right]\Big(^1_0 x$$

which is a continuous homogeneous polynomial of second degree in x. The result follows by the identity principle.

§6.- The pseudorotations on autD.

9.19. DEFINITION. *For* $t\in\mathbb{R}$, *we introduce the mappings* ϕ^t: $\mathbb{C}$ autD $\to$ $\mathbb{C}$ autD *in the following manner:*

(a) *Let* $A\in$autD *be given. Then* A *admits a unique representation* $A=A_c+L$ *with* $A_c\in Q$ *and* $L\in L$ *and we define*

$$\phi^t: A=A_c+L \to A_{e^{it}c}+L$$

(b) *Now* ϕ^t *may be extended to* $\mathbb{C}$autD *by complex linearity because we have* $\mathbb{C}$autD= (autD)$\oplus$i(autD), *the sum being direct.*

In order to show that ϕ^t is a Lie algebra automorphism of autD, we introduce an auxiliary transformation.

9.20. DEFINITION. *For* $t\in\mathbb{R}$, *we define the mapping* $R^t: E\to E$ *by* $R^t: x\to e^{it}x$.

Let $A\in$autD be given and denote by $J: V\to U$ the neighbourhood U of 0 and the isomorphism J given by definition 9.11. Thus, A is uniquely determined by its restriction to U and $J_{\#}\phi^t A$ is a holomorphic vector field on U= J(V). Besides, by proposition 9.18, $J_{\#}A$ is an entire holomorphic vector field (actually, $J_{\#}A$ is a polynomial of degree not greater than 2) so that $R^t_{\#}J_{\#}A$ is also an entire holomorphic vector field and it makes sense to compare $(R^t_{\#}J_{\#}A)_{|U}$ with $J_{\#}(\phi^t A)_{|U}$. We get the following result

9.21. PROPOSITION. (a) *We have* $R^t_{\#}J_{\#}A=J_{\#}\phi^t A$ *for all* $t\in\mathbb{R}$

and $A\in \mathrm{aut}D$.

(b) *For all* $t\in\mathbb{R}$, ϕ^t *is a Lie algebra automorphism of* autD.

Proof: (a) Since any $A\in \mathrm{aut}D$ may be written in the form $A= C_c+Q_c+L$ for some $c\in E$ and $L\in\mathcal{L}$, it suffices to check the equality $R^t_{\#}\, J_{\#}=J_{\#}\phi^t$ in these particular vector fields. Let $c\in E$ be given. By proposition 9.18, $J_{\#}C_c$ is constant; thus we have

$$(J_{\#}\phi^t C_c)x= J_{\#}(e^{it}C_c)x= e^{it}(J_{\#}C_c)x= e^{it}c$$

and

$$(R^t_{\#}J_{\#}C_c)x= (R^t)'_{R^{-t}x}J_{\#}C_c(R^{-t}x)= e^{it}(J_{\#}C_c)e^{-it}x= e^{it}c$$

As Q_c is a homogeneous polynomial of second degree, we have

$$(J_{\#}\phi^t Q_c)x= J_{\#}Q_{e^{it}c}x= J_{\#}(e^{-it}Q_c)x= e^{-it}(J_{\#}Q_c)x$$

and

$$(R^t_{\#}J_{\#}Q_c)x= (R^t)'_{R^{-t}x}J_{\#}Q_c(R^{-t}x)= e^{it}J_{\#}Q_c(e^{-it}x)= e^{it}e^{-2it}J_{\#}Q_c(x)=$$

$$= e^{-it}(J_{\#}Q_c)x$$

Since for $L\in\mathcal{L}$, $J_{\#}L$ is linear, we have

$$(J_{\#}\phi^t L)x= J_{\#}Lx= e^{it}J_{\#}Le^{-it}x= R^tJ_{\#}LR^{-t}x= (R^t_{\#}J_{\#}L)x$$

(b) Obviously, ϕ^t is an isomorphism of autD as a vector space. By step (a), for $A_1, A_2\in \mathrm{aut}D$, we have

$$\phi^t[A_1,A_2]_{|U}= J_{\#}^{-1}J_{\#}\phi^t[A_1,A_2]= J_{\#}^{-1}R^t_{\#}J_{\#}[A_1,A_2]= J_{\#}^{-1}[R^t_{\#}J_{\#}A_1, R^t_{\#}J_{\#}A_2]=$$

$$= J_{\#}^{-1}[J_{\#}\phi^t A_1,J_{\#}\phi^t A_2]= J_{\#}^{-1}J_{\#}[\phi^t A_1,\phi^t A_2]=$$

$$= [\phi^t A_1, \phi^t A_2]_{|U}$$

whence the conclusion follows by the identity principle.

#

Let us denote by π_j, j= 0,1, the canonical projections associated with the decomposition autD= $\mathcal{L}\oplus\mathcal{Q}$.

9.22. LEMMA. *The norm* $|\cdot|$ *defined on* autD *by*

$$|A| =: \max\{\|(\pi_j A)_0^{(k}\| ;\quad j,k= 0,1\}$$

is invariant under all transformations ϕ^t, $t\in\mathbb{R}$. *Moreover, it defines the natural topology on* autD.

Proof: Let us suppose that we have $A_n \to 0$ in autD; then $\pi_j A_n \to 0$ in autD, so that $\|\pi_j A_n\| = \sum_{k=0}^{1} \|(\pi_j A_n)_0^{(k}\| \to 0$; thus $|A_n| \to 0$. Conversely, if $|A_n| \to 0$ then $\|(\pi_j A_n)_0^{(k}\| \to 0$ for j,k= 0,1; thus by theorem 5.6, we have $\pi_j A_n \to 0$ (j= 0,1) and $A_n = \pi_0 A_n + \pi_1 A_n \to 0$.

Moreover, from $\phi^t A = \phi^t(A_c + L) = A_{e^{it}c} + L$ we get

$$\pi_0 \phi^t A = \phi^t \pi_0 A \quad , \quad \pi_1 \phi^t A = \phi^t \pi_1 A$$

for all $a\in$autD. From definition 9.4 we obtain $(A_{\lambda c})_0^{(k} = \lambda (A_c)_0^{(k}$ for $\lambda\in\mathbb{C}$, which completes the proof.

#

9.23. DEFINITION. *Let* $\mathcal{N}$ *be a neighbourhood of the origin in* autD *such that*

(a) *the mapping* $A\in\mathcal{N} \to \exp A\in$AutD *is injective*

(b) $\mathcal{N}$ *is invariant under all transformations* ϕ^t, $t\in\mathbb{R}$.

Then we set $G =: \exp\mathcal{N}$ *and define*

$$\psi^t : \exp A \to \exp \phi^t A$$

for $t\in\mathbb{R}$ and $A\in N$.

Observe that by lemma 6.47 and lemma 9.22 such a neighbourhood exists.

Our next task will be to extend the mappings $\Psi^t: G \to G$ to the identity component $\mathrm{Aut}_0 D$ of AutD. By lemma 7.15, any $G\in\mathrm{Aut}_0 D$ admits a representation of the form $G= G_1 G_2 .. G_n$ with $G_k= \exp A_k$ and $A_k \in N$ for $k= 1,2,..,n$ so that we could set

$$\Psi^t G=: (\Psi^t G_1)..(\Psi^t G_n)$$

The trouble is that the representation of G we have used is not unique.

9.24. PROPOSITION. *Let* $G_1, G_2, .., G_n \in G$ *be such that* $G_1 \circ G_2 \circ .. \circ G_n = \mathrm{id}_D$. *Then we have* $(\Psi^t G_1).(\Psi^t G_2)..(\Psi^t G_n)= \mathrm{id}_D$ *for all* $t\in\mathbb{R}$.

Proof: Let us write

$$G^t=: (\Psi^t G_1).(\Psi^t G_2)..(\Psi^t G_n)$$

for $t\in\mathbb{R}$. We begin with the following observation: Given any $X\in\mathrm{aut}D$, we have

$$G^t_{\#} X= X \tag{9.9}$$

for all $t\in\mathbb{R}$. Indeed, by assumption, there are $A_k \in N$, $k= 1,2,..,n$ such that $G_k= \exp A_k$. Write $X=: \phi^t Y$ where $Y=: \phi^{-t} X\in\mathrm{aut}D$ and t is kept fixed; then, by proposition 5.13 and lemma 5.14, we have

$$G^t_{\#} X= [(\Psi^t G_1)..(\Psi^t G_n)]_{\#} X= (\exp\phi^t A_1)_{\#} .. (\exp\phi^t A_n)_{\#} \phi^t Y=$$

$$= [\exp(\phi^t A_1)_{\#}]..[\exp(\phi^t A_n)_{\#}]\phi^t Y= \phi^t (\exp A_{1\#})..(\exp A_{n\#}) Y=$$

$$= \phi^t (G_{1\#} .. G_{n\#}) Y= \phi^t (G_1 .. G_n)_{\#} Y= \phi^t (\mathrm{id}_D)_{\#} Y= \phi^t Y= X.$$

From (9.9) we can deduce $G^t= \mathrm{id}_D$. Indeed, set

$$F^h =: G^{t+h}(G^t)^{-1}$$

for t, h∈ℝ. By (9.9) we have

$$F^h_\# X = X \tag{9.10}$$

for all X∈autD and h∈ℝ.

Now we show that the mapping h∈ℝ → F^h∈AutD is $\mathcal{T}$derivable at h= 0. Let A= A_c+L∈autD be fixed; since c∈E → A_c∈autD is a continuous real-linear mapping, we have

$$\frac{d}{dt}\phi^t A = \lim_{h\to 0}\frac{1}{h}(\phi^{t+h}A - \phi^t A) = \lim_{h\to 0}(A_{c\,e^{i(t+h)}} - A_{c\,e^{it}}) =$$

$$= A_{c\lim_{h\to 0}\frac{1}{h}(e^{i(t+h)}-e^{it})} = A_{i\,e^{it}c}$$

Moreover, by lemma 6.45, the mapping A→expA, A∈N, is real analytic with regard to the $\mathcal{T}$ topology on AutD. Thus, considering the composed mapping

$$t \to \phi^t A \to \exp\phi^t A\ ,$$

we get the $\mathcal{T}$ derivability of t → $\Psi^t H$ with H= expA. Applying this to each of the $\Psi^t G_k = \exp\phi^t A_k$, k= 1,2,...,n, by lemma 1.15 we get the $\mathcal{T}$weak derivability of G^t. Therefore, for some neighbourhood B of 0, we have

$$\frac{1}{h}(G^{t+h}-G^t) \text{ converges in the norm } \|\cdot\|_B$$

or, equivalently,

$$\frac{1}{h}(F^h - id_D) \text{ converges in the norm } \|\cdot\|_{G^t(B)}$$

whence it follows that

$$\mathcal{T}\lim_{h\to 0}\frac{1}{h}(F^h_\# - id_D) = A^t$$

for some $A^t \in \mathrm{aut}D$, so that $h \to F^h$ is T derivable at $h=0$.

Then, theorem 4.28 entails

$$T\lim_{h\to 0} \frac{1}{h}(F^h_{\#}X-X) = [A^t,X]$$

for all $X\in \mathrm{aut}D$, so that by (9.10) we have $[A^t,X]=0$ for all $X\in\mathrm{aut}D$. Thus, by proposition 9.13

$$0=A^t= T\lim_{h\to 0}\frac{1}{h}[G^{t+h}(G^t)^{-1}-G^t(G^t)^{-1}] = T\lim_{h\to 0}\frac{1}{h}(G^{t+h}-G^t)(G^t)^{-1}$$

whence $\frac{d}{dt}G^t=0$ for all $t\in\mathbb{R}$, that is, G^t is constant and $G^t=G^0=\mathrm{id}_D$.

#

9.25. COROLLARY. *Let* $G_1,G_2..G_n\in\mathcal{G}$ *and* $H_1,H_2..H_m\in\mathcal{G}$ *be given and assume that* $G_1\circ G_2\circ..\circ G_n = H_1\circ H_2\circ..\circ H_m$. *Then* $(\Psi^tG_1)..(\Psi^tG_n)=(\Psi^tH_1)..(\Psi^tH_m)$ *for all* $t\in\mathbb{R}$.

Proof: We need only to observe that, if $H=\exp A$ with $A\in\mathcal{N}$, then

$$(\Psi^tH)^{-1} = (\exp\phi^tA)^{-1} = \exp(-\phi^tA) = \exp[\phi^t(-A)] =$$

$$= \Psi^t\exp(-A) = \psi^t(H^{-1}).$$

#

9.26. DEFINITION. *For* $t\in\mathbb{R}$, *we define the mapping* $\Psi^t\colon \mathrm{Aut}_0D\to\mathrm{Aut}_0D$ *by means of*

$$\Psi^t\colon G_1..G_n \to (\Psi^tG_1)..(\Psi^tG_n)$$

whenever $G_1,..,G_n\in\mathcal{G}$.

We know that $\mathrm{Aut}_0D=\bigcup_{n\in\mathbb{N}}\mathcal{G}^n$; therefore, in view of the previous proposition, the mappings Ψ^t are well-defined on Aut_0D. Moreover, we have

$$(\Psi^t G)(\Psi^{-t} G) = G \quad \text{and} \quad \Psi^t(GH) = (\Psi^t G)(\Psi^t H),$$

for $G, H \in \mathrm{Aut}_0 D$ and $t \in \mathbb{R}$.

9.27. EXERCISES. Consider the mapping $R \times \mathrm{Aut}_0 D \to \mathrm{Aut}_0 D$ given by $(t,G) \to \Psi^t G$. Show that

(a) For fixed G, the application $(t,G) \to \Psi^t G$ is a one-parameter group

(b) The joint application $(t,G) \to \Psi^t G$ is real analytic when $\mathrm{Aut}_0 D$ is endowed with the analytic topology $\mathcal{T}_a$. Is it $\mathcal{T}$ continuous?.

§7.- The pseudorotations on D.

We recall that, by proposition 9.8, D is homogeneous under the action of $\mathrm{Aut}_0 D$, so that

$$D = \{G(0); \quad G \in \mathrm{Aut}_0 D\}$$

9.28. DEFINITION. *For* $t \in \mathbb{R}$, *we define* $T^t : D \to D$ *in the following manner: Let* $x \in D$ *be given; then we have* $x = G(0)$ *for some* $G \in \mathrm{Aut}_0 D$, *and we set*

$$T^t x =: (\Psi^t G) 0$$

In order to see that this definition makes sense we have to verify that, for $G_1, G_2 \in \mathrm{Aut}\, D$ with $G_1 0 = G_2 0$, we have $(\Psi^t G_1) 0 = (\Psi^t G_2) 0$. By passing to $G =: G_1^{-1} G_2$ we must prove that, for all $G \in \mathrm{Aut}_0 D$, the relation $G0 = 0$ implies $(\Psi^t G) 0 = 0$ for all $t \in \mathbb{R}$. This will be our next task.

9.29. DEFINITION. *We set*

$$\mathrm{Isot} D =: \{G \in \mathrm{Aut}\, D; \quad G0 = 0\}$$

So far, we have made no use of the assumption concerning the *simple connectivity* of D. We shall apply it to prove the following:

9.30. PROPOSITION. *Assume that the bounded symmetric domain* D *is simply connected. Then the subgroup* IsotD *is arcwise connected with regard to the topology* T_a.

Proof: It suffices to show that, for any $G \in \text{IsotD}$, there exists a T_a continuous path Γ: $[0,1] \to \text{IsotD}$ such that $\Gamma(0) = id_D$ and $\Gamma(1) = G$.

Let $G \in \text{IsotD}$ be given; then we have

$$G \in \text{Aut } D \qquad \text{and} \qquad G0 = 0.$$

Therefore, we can find $A_1, A_2, \ldots, A_n \in N$ such that

$$G = (\exp A_1) \circ (\exp A_2) \circ \ldots \circ (\exp A_n)$$

We divide the interval $[0,1]$ into n subintervals

$$[0, \frac{1}{n}], \ [\frac{1}{n}, \frac{2}{n}] \ldots [\frac{n-1}{n}, 1]$$

and define the auxiliary mapping $\hat{\Gamma}$: $[0,1] \to \text{Aut}_0 D$ by means of

$$G_t =: (\exp A_1) \circ (\exp A_2) \circ \ldots \circ (\exp A_k) \exp[n(t - \frac{k}{n}) A_{k+1}]$$

for $t \in [\frac{k}{n}, \frac{k+1}{n}]$ and $k = 0,1,\ldots,n-1$. Obviously, $\hat{\Gamma}$ is a T_a continuous path which connects id_D and G in the space $\text{Aut}_0 D$. In order to connect them in the subspace IsotD, we project this path $\hat{\Gamma}$: $[0,1] \to \text{Aut}_0 D$ into D by applying each $\hat{\Gamma}(t) = G_t$ to the origin 0, so that we get the path γ: $[0,1] \to D$ defined by

$$\gamma(t) =: G_t(0)$$

Since G belongs to IsotD, γ is a closed path: $\gamma(0) = id_D(0) = 0$ and $\gamma(1) = G(0)$. Thus, as D is assumed to be simply connected, γ is homotopic to the origin 0. Let us denote by $R =: [0,1] \times [0,1]$ the unit rectangle and denote by f: $(s,t) \in R \to f(x,t) \in D$ a homotopy in D continuously deforming the path γ into the origin 0, so that we have

$$(9.11)\qquad \begin{array}{ll} f(0,t)=\gamma(t)=G_t(0) & f(1,t)=\mathrm{id}_D(0)=0 \\ f(s,0)=\mathrm{id}_D(0)=0 & f(s,1)=G(0)=0. \end{array}$$

We shall construct a lifting of f: R → D to $\mathrm{Aut}_0 D$, i.e., a T_a continuous function f: R → $\mathrm{Aut}_0 D$ such that

$$F(s,t)0=f(s,t)$$

for all (s,t)∈R. Then we shall have F(1,t)0= f(1,t)= 0 for all t∈[0,1], so that F(1,t)∈IsotD for t∈[0,1] and, by writing

$$\Gamma =: t \to F(1,t)$$

we obtain the path Γ: [0,1] → IsotD we were looking for.

Let U be the neighbourhood of the origin in E constructed in the proof of proposition 9.8; thus the mapping

$$(9.12)\qquad g: c\in U \to (\exp A_0)0 \in g(U)$$

is an isomorphism and

$$\psi(U)=\{A_c;\ c\in U\}\subset N$$

Then, we have

$$g(U)\supset B_\varepsilon(0)$$

for some carathéodorian open ball $B_\varepsilon(0)$ centered at 0. As the homotopy f: R → D is continuous, the mapping

$$[(s,t),\ (s',t')] \to d_D[f(s\ t);\ f(s',t')]$$

is uniformly continuous on R×R; therefore, there exists an m∈ℕ such that

$$(9.13)\quad (s,t),\ (s',t')\in \mathbb{R}\quad |s-s'|\leqslant \frac{1}{m}\qquad |t-t'|\leqslant \frac{1}{m} \Rightarrow d_D[f(x,t);f(s',t')]<\varepsilon$$

Now we devide the horizontal side $[0,1]$ of the rectangle R into m subintervals

$$[0, \frac{1}{m}], [\frac{1}{m}, \frac{2}{m}], \ldots, [\frac{m-1}{m}, 1]$$

and construct recurrently the lifting F of f on each of the subrectangles $R_k =: [\frac{k}{m}, \frac{k+1}{m}] \times [0,1]$, $k = 0,1,\ldots,m-1$.

We claim that, for $(s,t) \in R_1$, we have

$$G_t^{-1} f(s,t) \in B_\varepsilon(0)$$

Indeed, as the carathéodorian distance is AutD-invariant, by (9.11) we have

$$d_D[G_t^{-1} f(s,t), 0] = d_D[f(s,t), G_t 0] = d_D[f(s,t), f(0,t)] < \varepsilon$$

Therefore, by (9.12) it makes sense to apply J^{-1} to $G_t^{-1} f(s,t)$ and we define

$$c(s,t) =: J^{-1} G_t^{-1} f(s,t)$$

for $(s,t) \in R_0$. Let us set

$$F_0(s,t) =: G_t \exp A_{c(s,t)}$$

for $(s,t) \in R_0$. Then, it is easy to check that F_0 is a lifting of f over R_0.

Now we proceed by induction on k. Assume we had already constructed a lifting F_k of f over R ; thus

$$F_k(s,t)0 = f(s,t)$$

for all $(s,t) \in R_k$. We claim that, for $(s,t) \in R_{k+1}$, we have

$$F_k^{-1}(\frac{k}{m}, t) f(s,t) \in B_\varepsilon(0)$$

Indeed, by (9.13) and the induction hypothesis we have

$$d_D[F_k^{-1}(\frac{k}{m},t)f(s,t),0] = d_D[f(s,t),F_k(\frac{k}{m},t)0] =$$

$$= d_D[f(s,t),f(\frac{k}{m},t)] < \varepsilon$$

Thus, it makes sense to apply J^{-1} to $F_k^{-1}(\frac{k}{m},t)f(s,t)$ and we define

$$c(s,t) =: J^{-1}F_k^{-1}(\frac{k}{m},t)f(s,t)$$

for $(s,t)\in R_{k+1}$. If we set

$$F_{k+1}(s,t) =: F_k(\frac{k}{m},t)\exp A_{c(s,t)}$$

for $(s,t)\in R_{k+1}$, then it is easy to check that F_{k+1} lifts f on R_{k+1}. Moreover, F_{k+1} and F_k agree on the common border of their rectangles of definition:

$$F_{k+1}(\frac{k}{m}, t) = F_k(\frac{k}{m},t) \qquad \forall t\in[0,1]$$

so that F_{k+1} extends the previous partial lifting. This completes the proof.

#

Let U and V be the neighbourhoods of 0 in E constructed in definition 9.8 and put

$$R^t =: x \to e^{it}x$$

for $t\in\mathbb{R}$ and $x\in E$. We may assume that U is an open ball centered at 0, so that U is invariant under the transformations R^t. By setting

$$S^t =: J\circ R^t\circ J^{-1}$$

it is easy to see that V is invariant under the transformations S^t. Finally, we recall that, by proposition 9.21, we have

(9.14) $$\phi^t A_{|V} = (J_{\#}^{-1} \circ R_{\#}^{t} \circ J\ ) A_{|V} = S_{\#}^{t} A_{|V}$$

for all $A \in \mathrm{aut} D$.

9.31. LEMMA. *Let* $t \in \mathbb{R}$ *be given. Then, there are a number* $\delta > 0$ *and a neighbourhood* W *of 0 such that we have*

$$(\Psi^t \exp A)_{|W} = [S^t (\exp A) S^{-t}]_{|W}$$

for all $A \in \mathrm{aut} D$ *with* $\| A \|_V < \delta$.

Proof: Take any $\delta > 0$ such that the ball $B_{2\delta}(0)$ with center at 0 and radius 2δ is contained in V; then $S^t[B_\delta(0)] \subset V$ is a neighbourhood of 0 and we define

$$W =: B_\delta(0) \cap S^t[B_\delta(0)]$$

The pair δ, W satisfies our requirements.

Indeed: Let $A \in \mathrm{aut} D$ be such that $\| A \|_V < \delta$ and take any $x \in B_\delta(0)$. Consider the initial value problem

$$\frac{d}{dt} y(t) = A[y(t)], \quad y(0) = x$$

whose solution is denoted by $y(t) = (\exp tA)x$, and set

$$\tau(x) =: \inf\{t > 0;\ \| (\exp tA)x - x \| \geq \delta\}$$

We claim that $\tau(x) \geq 1$. Indeed, for $0 \leq t < \tau(x)$ we have

$$\| (\exp tA)x - x \| = \| \int_0^t \frac{d}{ds} y(s) ds \| = \| \int_0^t A[(\exp sA)x] ds \| \leq$$

$$\leq \int_0^t \| A \|_{B_{2\delta}(0)} ds \leq \int_0^t \delta ds = \delta t < \delta \tau(x)$$

If it were $\tau(x) < 1$ for some $x \in B_\delta(0)$, by taking α with $\tau(x) < \alpha < 1$, we would obtain

$$\|(\exp tA)x-x\| \leqslant \delta\alpha < \delta$$

for all $t \in [0, \tau(x))$, so that

$$\inf\{t>0;\ \|(\exp tA)x-x\| \geqslant \delta\} > \tau(x)$$

which is contradictory. Thus, $\tau(x) \geqslant 1$ for all $x \in B_\delta(0)$ and we have

$$(\exp A)W \subset (\exp A)B_\delta(0) \subset B_{2\delta}(0) \subset V$$

Therefore, it makes sense to apply $S^t(\exp A)S^{-t}$ to W and, by (9.14), we get

$$[S^t(\exp A)S^{-t}]W = (\exp S^t_\# A)W = (\exp \phi^t A)W = (\Psi^t \exp A)W.$$

#

<u>9.32. PROPOSITION</u>. *We have* $(\Psi^t G)0 = 0$ *for all* $G \in \mathrm{Isot}D$ *and all* $t \in \mathbb{R}$.

Proof: We recall that the family

$$\{\exp A;\ \|A\|_V < \frac{1}{n},\ n \in \mathbb{N}\}$$

is a fundamental system of $\mathcal{T}_a$-neighbourhoods of id_D in $\mathrm{Aut}D$; therefore, the family of subsets

$$\mathcal{G}_n =: \{(H_1, H_2);\ H_1^{-1}H_2 = \exp A,\ \|A\|_V < \frac{1}{n}\}$$

is a bases for the left uniform structure on the topological group $(\mathrm{Aut}D, \mathcal{T}_a)$.

Now, let $G \in \mathrm{Isot}D$ and $t \in \mathbb{R}$ be given. By proposition 9.30 there is a $\mathcal{T}_a$ continuous path $\Gamma: [0,1] \to \mathrm{Isot}D$ connecting id_D and G in $\mathrm{Isot}D$. Let δ and W be determined as in lemma 9.31. Since Γ is uniformly continuous on $[0,1]$, there is a partition $0 = s_0 < s_1 < \ldots < s_{m-1} < s_m = 1$ of $[0,1]$ such that we have

$$G_{s_j}^{-1}\, G_{s_{j+1}} \in \{\exp A;\ \|A\|_V < \delta\}$$

and, by lemma 9.31

$$\Psi^t(G_{s_j}^{-1} G_{s_{j+1}})|_W = [S^t(G_{s_j}^{-1} G_{s_{j+1}})S^{-1}]|_W$$

for all j= 0,1,..,m. Since $0 \in W$ and $G_{s_j} \in \text{Isot}D$, we get

$$\Psi^t(G_{s_j}^{-1} G_{s_{j+1}})0 = [S^t(G_{s_j}^{-1} G_{s_{j+1}})S^{-t}]0 = 0$$

for all j, and finally

$$(\Psi^t G)0 = \Psi^t[(G_0^{-1} G_1)(G_1^{-1} G_2)..(G_{m-1}^{-1} G_m)] =$$

$$= \Psi^t(G_0^{-1} G_1)\Psi^t(G_1^{-1} G_2)..\Psi^t(G_{m-1}^{-1} G_m)0 = 0.$$

#

<u>9.33. PROPOSITION</u>. *The mappings* T^t: D → D *of definition 9.28 are well defined on* D. *Moreover, for all* $t \in \mathbb{R}$, T^t *is a holomorphic extension of* $S^t = J^{-1} \circ R^t \circ J$ *and* $t \to T^t$ *is a T-continuous one-parameter group* $\mathbb{R} \to \text{Aut}_0 D$.

Proof: We have already shown that the T^t are well defined on D. Now we prove that they are holomorphic in D. Let $x \in D$ be given and fix any $G \in \text{Aut}_0 D$ with G0= x. Since J: $c \to (\exp A_c)0$ is real bianalytic in some neighbourhood V of 0, so are the mappings

$$F_1 =: c \to G(\exp A_c)0$$

$$F_2 =: c \to \Psi^t(G\exp A_c)0 = (\Psi^t G)(\exp A_{e^{it}c})0$$

for t fixed. Therefore $T^t|_{F_1(V)} = F_2 F_1^{-1}$ is real analytic in a neighbourhood $F_1(V)$ of x. By the arbitrariness of x, T^t is real analytic in D. In particular, for x= 0, G= id_D and V= W (where W is the neighbourhood of 0 constructed in lemma 9.31), we have

$$T^t(\exp A_c)0 = (\Psi^t \exp A_c)0 = [S^t(\exp A_c)S^{-1}]0 = S^t(\exp A_c)0$$

for all $c \in W$. This means that T^t is an extension of S^t. Since T^t is real analytic in D and, in W, T^t coincides with S^t, which is holomorphic, T^t is holomorphic in D.

Obviously, we have $T^t(D) \subset D$ for all $t \in \mathbb{R}$. Moreover, from the definition of T^t, we can easily obtain $T^{s+t} = T^s T^t$ for all $s,t \in \mathbb{R}$; thus $T^t \in \mathrm{Aut}_0 D$ and $t \to T^t$ is a one-parameter group in $\mathrm{Aut}_0 D$. To show its $\mathcal{T}$-continuity it suffices to prove that

$$\mathcal{T} \lim_{t \to 0} T^t = id_D \tag{9.15}$$

Since $T^t{}_{|V} = S^t$, we have

$$\lim_{t\to 0} T^t = \lim_{t\to 0} S^t = \lim_{t\to 0} J^{-1} \circ R^t \circ J = id_D \quad \text{uniformly on } V.$$

As the transformations T^t are automorphisms of D, this entails (9.15) by theorem 1.6.

§8.- Construction of the image domain.

Next, a bounded balanced domain $\hat{D}$, which is biholomorphically equivalent to D, can be defined in terms of the transformations T^t.

9.34. DEFINITION. *We introduce the mapping* $F: D \to E$ *by*

$$F =: x \to \frac{1}{2\pi} \int_0^{\pi} e^{-it}\, T^t(x)\,dt$$

and we set

$$\hat{D} =: F(D).$$

9.35. LEMMA. $\hat{D}$ *is bounded, and* F *is a holomorphic extension of* J.

Proof: We have

$$\sup_{y\in\hat{D}} \| y\| \leqslant \sup_{x\in D} \frac{1}{2\pi}\int_0^{2\pi} \| T^t(x)\| \, dt \leqslant \sup_{x\in D}\| x\| <\infty$$

The holomorphy of F is clear since $x \to T^t(x)$ is holomorphic for all $t\in\mathbb{R}$ and $t \to T^t$ is $\mathcal{T}$continuous.

Let $x\in V$ be given. As $T^t{}_{|V} = S^t = J^{-1}\circ R^t\circ J$ and V is invariant under the transformations S^t, we have

$$F(x) = \frac{1}{2\pi}\int_0^{2\pi} e^{-it}\ T^t(x)dx = \frac{1}{2\pi}\int_0^{2\pi} e^{-it}\ J^{-1}\ R^t\ J(x)dt =$$

$$= \frac{1}{2\pi}\int_0^{2\pi} e^{-it}\ J^{-1}(e^{it}J(x))dt = \frac{1}{2\pi i}\int_{|\zeta|=1} \frac{1}{\zeta^2}\ J^{-1}[\zeta J(x)]d\zeta =$$

$$= \frac{d}{d\zeta}\Big|_0\ J^{-1}(\zeta J(x)) = (J^{-1})_0^{(1}J(x) = J(x)$$

#

9.36. COROLLARY. *Given* $A\in\mathbb{C}autD$, *there is a unique polynomial* $\hat{A}$: $E \to E$ *of degree two such that we have* $F^{(1}A = \hat{A}F$.

Proof: Let $A\in\mathbb{C}autD$ be given. By proposition 9.18, the mapping $\hat{A} =: J_{\#}A$ is a polynomial of degree two that satisfies

$$\hat{A}x = (J_{\#}A)x = J^{(1}_{J^{-1}x}\ A[J^{-1}(x)] = F^{(1}_{F^{-1}(x)}\ A(F^{-1}(x))$$

in the neighbourhood U of 0; thus, we have $AF = F^{(1}A$ in D. Any other $\hat{A}'\in\mathbb{C}autD$ satisfying $F^{(1}A = \hat{A}'J$ coincides with $J_{\#}A$ on U; therefore $\hat{A}$ is uniquely determined.

#

9.37. DEFINITION. *For* $A\in\mathbb{C}autD$, $\hat{A}$ *denotes the unique polynomial of degree two associated with* A *by corollary 9.36.*

From proposition 9.18, we know that the polynomials $\hat{C}_c$, $\hat{Q}_c$ and $\hat{L}$ are given by

$$\hat{C}_c(x) = c, \quad \hat{Q}_c(x) = \frac{1}{2}\,[Q_c, C_x]_0^{(1}x\ , \quad \hat{L}(x) = L_0^{(1}x$$

whatever are $c\in E$ and $L\in\mathcal{L}$.

9.38. PROPOSITION. $\hat{D}$ *is a connected open circular set.*

Proof: Obviously, $0\in\hat{D}$. Since $F_{|V} = J$ and $T^t{}_{|V} = S^t$, we have

$$FT^t(x) = FS^t(x) = J(J^{-1}\ R^tJ)(x) = e^{it}J(x) = e^{it}F(x)$$

for all $x\in V$ and $t\in\mathbb{R}$. As F and T^t are holomorphic on D, from the identity principle we derive

$$FT^t = e^{it}F$$

for all $t\in\mathbb{R}$. Then

$$e^{it}\hat{D} = e^{it}F(D) = FT^t(D)\ F(D) = \hat{D}$$

so that $\hat{D}$ is circular.

Next, let $x\in D$ be given and denote by B any ball centered at x with radius $\delta<\text{dist}(x,\delta D)$. Since the mapping $c\to C_c$ is continuous, there exists $\varepsilon>0$ such that we have $\|C_c\|_B<\delta$ whenever $\|c\|<\varepsilon$. We claim that the initial value problem

$$\frac{d}{dt}\,y(t) = C_c[y(t)], \qquad y(0) = x \tag{9.16}$$

has a maximal solution defined on the interval $[0,1]$ whatever is $c\in E$ with $\|c\|<\varepsilon$.

Indeed, denote by y_c the solution of (9.16) and set

$$\tau(c) =: \sup\{t>0;\ \ y_c(s)\in B \quad \text{for all } s\in[0,t]\}$$

Assume that for some $c = \xi$ we had $\tau(\xi)<1$. Then

$$\overline{\lim_{t\uparrow\tau(\xi)}}\ \|y_\xi(t)-x\| = \delta$$

Moreover, for $t\leqslant\tau(\xi)$, we have

$$\| y_\xi(t)-x\| \leqslant\int_0^t \| \frac{d}{ds} y_\xi(s)\| ds\leqslant\int_0^t \delta ds= \delta t< \delta\tau(\xi)$$

so that

$$\overline{\lim_{t\uparrow\tau(\xi)}} \| y_\xi(t)-x\| \leqslant\delta\tau(\xi)<\delta$$

which is a contradiction.

Therefore, for each $c\in E$ with $\| c\| <\varepsilon$, we can define a curve $\hat{y}_c: [0,1] \to \hat{D}$ by means of

$$\hat{y}_c(t)=: F(y_c(t))$$

and we have

$$\frac{d}{dt} y_c(t)= F^{(1}_{y_c(t)} \frac{d}{dt} y_c^{(t)} = F^{(1}_{y_c(t)} C_c(y_c(t))= \hat{C}_c[Fy_c(t)]= c$$

for all $t\in[0,1]$ and $c\in E$ with $\| c\| <\varepsilon$. But then

$$\hat{y}_c(t)= tc+F(x)$$

so that

$$\hat{D}= F(D)\supset\{c+F(x);\ \| c\| <\varepsilon\}$$

which is a neighbourhood of $F(x)$. As x was arbitrary in D, $\hat{D}$ is open. Obviously, $\hat{D}= F(D)$ is connected.

#

§9.- <u>The isomorphism between the domains D and $\hat{D}$</u>.

We complete the proof of Vigué's theorem by establishing that $\hat{D}$ is balanced and that the mapping $F: D \to \hat{D}$ is injective.

As we already know that $\hat{D}$ is a bounded circular domain, we may speak of $\mathrm{aut}\hat{D}$ and $\mathrm{Aut}\hat{D}$.

9.39. LEMMA. *We have* $\hat{A}\in\mathrm{aut}\hat{D}$ *for all* $A\in\mathrm{aut}D$. *Furthermore, for every* $G\in\mathrm{Aut}_0D$, *there is a unique* $\hat{A}\in\mathrm{Aut}_0\hat{D}$ *such that* $FG=\hat{G}F$.

Proof: Fix $A\in\mathrm{aut}D$ and $x\in D$ arbitrarily, and set

$$y(t)=:(\exp tA)x, \qquad \hat{y}(t)=Fy(t)$$

for $t\in\mathbb{R}$. Then we have

$$\frac{d}{dt}\hat{y}(t)=F^{(1}_{y(t)}\frac{d}{dt}y(t)=F^{(1}_{y(t)}A(y(t))=\hat{A}F(y(t))=\hat{A}(\hat{y}(t))$$

for all $t\in\mathbb{R}$; thus, if we put $\hat{x}=:Fx$, the curve $t\to\hat{y}(t)$ is the solution of the initial value problem

$$\frac{d}{dt}\hat{y}(t)=\hat{A}(\hat{y}(t)) \qquad \hat{y}(0)=\hat{x}$$

that is,

$$\hat{y}(t)=(\exp t\hat{A})\hat{x}$$

Thus, the vector field $\hat{A}$ is complete in $\hat{D}$, i.e., $\hat{A}\in\mathrm{aut}\hat{D}$, and we have

$$F(\exp tA)x=Fy(t)=\hat{y}(t)=(\exp t\hat{A})\hat{x}=(\exp t\hat{A})Fx)$$

for all $x\in D$ and $t\in\mathbb{R}$. Therefore

$$F\exp A=\exp\hat{A}F.$$

Now, let $G\in\mathrm{Aut}_0D$ be given. Then we can write $G=(\exp A_1)..(\exp A_n)$ for some $A_1,..,A_n\in\mathrm{aut}D$, so that

$$FG=F[(\exp A_1)..(\exp A_n)]=(\exp\hat{A}_1)F[(\exp A_2)..(\exp A_n)]=$$
$$=..=(\exp\hat{A}_1)..(\exp\hat{A}_n)F=\hat{G}F$$

where $\hat{G}=:(\exp\hat{A}_1)..(\exp\hat{A}_n)\in\mathrm{Aut}\hat{D}$. Besides, since $F_{|V}=J$, from $\hat{G}F=FG$ we derive

$$\hat{G}_{|U} = (FGF^{-1})_{|U} = JGJ^{-1}$$

so that $\hat{G}$ is uniquely determined by G.

#

9.40. DEFINITION. *For* $G\in \mathrm{Aut}_0 D$ *we denote by* $\hat{G}$ *the unique element of* $\mathrm{Aut}_0\hat{D}$ *such that* $FG = \hat{G}F$.

9.41. COROLLARY. $\hat{D}$ *is a homogeneous balanced domain.*

Proof: From $F(0) = 0$ we derive $\hat{G}(0) = \hat{G}F(0) = FG(0)$ and

$$(\mathrm{Aut}_0\hat{D})0 = \{\hat{G}(0);\ G\in\mathrm{Aut}_0\hat{D}\} = \{FG(0);\ G\in\mathrm{Aut}_0 D\} =$$

$$= F\{G(0);\ G\in\mathrm{Aut}_0 D\} = F\left[(\mathrm{Aut}_0 D)0\right] = F(D) = \hat{D}$$

so that $\hat{D}$ is homogeneous. From corollary 7.10 we know that any homogeneous bounded circular domain $\hat{D}$ is balanced.

Let δ_D and $\delta_{\hat{D}}$ denote the carathéodorian differential metrics on D and $\hat{D}$ (cf. definition 3.8), respectively

9.42. LEMMA. *There is a constant* $M>0$ *such that we have*

$$\delta_D(x,v) \leqslant M\delta_{\hat{D}}(Fx, F_x^{(1}v)$$

for all $x\in D$, $v\in E$.

Proof: Let $x\in D$ and $v\in E$ be given. Fix any $G\in\mathrm{Aut}_0 D$ with $x = G0$ and put $u =: (G_0^{(1})^{-1}v$; by corollary 3.11 we have $\delta_D(0,u) = \delta_D(G0, G_0^{(1}u)$ whence

$$\delta_D(x,v) = \delta_D\left[0, (G_0^{(1})^{-1}v\right]$$

Moreover, we have $F0 = 0$ and $F_0^{(1} = \mathrm{id}$; thus, the relation $FG = \hat{G}F$ entails $Fx = G0$ and $F_x^{(1}\hat{G}_0^{(1} = \hat{G}_0^{(1}$. Therefore, again by corollary 3.11

$$\delta_{\hat{D}}(0,u) = \delta_{\hat{D}}(\hat{G}0,\ \hat{G}_0^{(1}u) = \delta_{\hat{D}}(Fx,\ F_x^{(1}G_0^{(1}u) = \delta_{\hat{D}}(Fx, F_x^{(1}v)$$

whence

$$\delta_{\hat{D}}(Fx, F_x^{(1}v) = \delta_{\hat{D}}[0, (G_0^{(1)})^{-1}v]$$

From definition 3.8 and the fact that $J = F_{|V}$ is an isomorphism between V and U, it is easy to derive

$$\delta_D(x,v) = \delta_D[0,(G_0^{(1)})^{-1}v] \leq \delta_V[0,(G_0^{(1)})^{-1}v] = \delta_U[F0, F_0^{(1}(G_0^{(1)})^{-1}v] =$$

$$= \delta_U[0,(G_0^{(1)})^{-1}v] \leq M\delta_{\hat{D}}[0,(G_0^{(1)})^{-1}v] = M\delta_{\hat{D}}(Fx, F_x^{(1}v)$$

for some $M>0$ independant of x and v.

#

<u>9.43. COROLLARY</u>. *The mapping* $c \to (\exp C_c)$ *is well defined for all* $c \in \hat{D}$.

Proof: First, we consider the mapping $c \to (\exp\hat{C}_c)0$. It is easy to see that the maximal solution of the initial value problem

$$\frac{d}{dt}\hat{y}(t) = \hat{C}_c[\hat{y}(t)], \quad \hat{y}(0) = 0$$

is given by $\hat{y}_c(t) = tc$. As $\hat{D}$ is balanced, the segment $[0,1]c$ is contained in $\hat{D}$ whatever is $c \in \hat{D}$. Therefore, $(\exp\hat{C}_c)0$ is well defined for all $c \in \hat{D}$.

Consider now the maximal solution y_c of

$$\frac{d}{dt}y(t) = C_c[y(t)], \quad y(0) = 0$$

for $c \in \hat{D}$ and set

$$\tau(c) =: \sup\{t>0;\ y_c(s) \in D \quad \text{for all } s \in [0,t]\}$$

We claim that $\tau(c) \geq 1$ for all $c \in \hat{D}$, so that $(\exp C_c)0$ is well defined for all $c \in \hat{D}$.

Indeed, assume we had $\tau(c) < 1$ for some $c \in \hat{D}$. Then, there does not exist any $a \in D$ with

$$\lim_{t\to\tau(c)} y_c(t) = a$$

By lemma 3.9, and lemma 9.42, for $0<t_1<t_2<\tau(c)$ we have:

$$(9.17)\quad d_D(y_c(t_1),\, y_c(t_2)) \leqslant \int_{t_1}^{t_2} \delta_D\left[y_c(t),\, \frac{d}{dt}\, y_c(t)\right]dt =$$

$$= \int_{t_1}^{t_2} \delta_D\left[y_c(t), C_c(y_c(t))\right]dt \leqslant M\int_{t_1}^{t_2} \delta_{\hat{D}}\left[Fy_c(t), F^{(1}_{y_c(t)} C_c(y_c(t))\right]dt =$$

$$= M\int_{t_1}^{t_2} \delta_{\hat{D}}\left[\hat{y}_c(t), \hat{C}_c(\hat{y}_c(t))\right]dt = M\int_{t_1}^{t_2} \delta_{\hat{D}}(tc,c)\,dt$$

As we have seen in the previous step, the segment $[0,1]c$ is contained in $\hat{D}$; therefore, for $t\in[t_1,t_2]$ we can find a ball $B(t)$ centered at tc and contained in $\hat{D}$.

By proposition 3.12 we have

$$\delta_{\hat{D}}(tc,c) \leqslant \delta_{B(t)}(Tc,c) \leqslant \frac{\|c\|}{\text{dist}(tc,\partial\hat{D})}$$

From the assumption $\tau(c)<1$ we derive $[t_1,t_2]c\subset[0,1]c\subset\hat{D}$, so that

$$(9.18)\quad \frac{\|c\|}{\text{dist}(tc,\partial\hat{D})} \leqslant \frac{\|c\|}{\text{dist}([0,1]c,\partial\hat{D})}$$

From (9.17) and (9.18) we finally get

$$d_D(y_c(t_1),y_c(t_2)) \leqslant \frac{M\|c\|}{\text{dist}([0,1]c,\partial\hat{D})}\,(t_2-t_1) \to 0$$

for $t_1,t_2 \nearrow t(C)$. However, as D is homogeneous, by theorem 3.18, D is complete with respect to the metric d_D; thus, there exists some $a\in D$ such that $\lim_{t\to\tau(c)} y_c(t) = a$, which is a contradiction.

#

Now, Vigué's theorem is at hand.

9.44. PROPOSITION. *The map* $F: D \to \hat{D}$ *is biholomorphic.*

Proof: It suffices to show that F is injective.

Now, $c \to C_c$ is a complex linear continuous mapping; therefore, it is holomorphic. From the theory of ordinary differential equations we know that, for $c \in \hat{D}$, the solution of the initial value problem

$$\frac{d}{dt} y(t) = C_c[y(t)], \quad y(0) = 0 \tag{9.19}$$

depends holomorphically on the parameter c. By corollary 9.43 the maximal solution of (9.19) is well defined on $[0,1]$ whatever is $c \in \hat{D}$, i.e.

$$\hat{F} =: c \to (\exp C_c)0$$

is a holomorphic mapping on $\hat{D}$. By lemma 9.35, we have $F_{|V} = J$ where $J: (\exp C_c)0 \to c$. Therefore we get $\hat{F}F = id_V$, and by the identity principle $\hat{F}F = id_D$. Similarly $F\hat{F} = id_{\hat{D}}$.

#

CHAPTER 10

THE JORDAN THEORY OF BOUNDED SYMMETRIC DOMAINS

§1.- Jordan triple product star algebras.

10.1. DEFINITION. Let $L_s(E\times E|E)$ denote the Banach space of all continuous symmetric bilinear mappings $E\times E \to E$ equipped with its usual norm and denote by $*: E \to L_s(E\times E|E)$ a continuous conjugate linear application $c \to c^*$. For $c\in E$ and $x,y\in E$ we write $xc^*y =: c^*(x,y)$. *Thus, for each fixed* $c\in E$, (E,c^*) *can be viewed as a "non necessarily associative" Banach algebra and the operation* $c^*: E\times E \to E$ *given by* $(x,y) \to xc^*y$ *is called the "*c^**-multiplication".*

The operation $E\times E\times E \to E$ *given by* $(x,c,y) \to xc^*y$ *is called the "triple product".* Clearly we have

$$\| xc^*y\| \leq M\| x\| \; \| c\| \; \| y\| \qquad x,c,y\in E$$

where M is the norm of the map $c \to c^*$.

The structure $(E,*)$ *is called a "Jordan triple product star algebra", or simply a* J^**-triple, if it satisfies the axioms:*

(J_1): $\quad x(a\,a^*b)^*x = 2(aa^*x)b^*x - aa^*(xb^*x)$

(J_2): $\quad (xa^*x)b^*x = xa^*(xb^*x)$

for every $a,b,x\in E$.

10.2. EXAMPLE. Let D be a bounded balanced circular domain in a Banach space E and $E_0 =: (\text{aut } D)0$. For $c\in E_0$, let $Q_c\in L_s(E\times E|E)$ denote the unique symmetric bilinear mapping such that the vector field $x \to c - Q_c(x,c)$, $x\in D$, belongs to aut D.

Then, according to proposition 7.9, the triple star product

$$xc^*y =: Q_c(x,y), \quad x,y \in E_0$$

fulfills (J_1) and (J_2), i.e. $(E_0,*)$ is a J^*-triple.

In particular, by Vigué's theorem, to every bounded symmetric domain D we may associate a J^*-triple on its supporting space in a natural way.

10.3. PROBLEM. How can be algebraically characterized those J^*-triples $(E,*)$ for which there exists some bounded balanced domain $D \subset E$ such that the vector fields of the form $x \to c - xc^*x$, $x \in D$, belong to aut D for all $c \in E$?.

In this chapter, our main purpose will be to answer this question.

In the sequel, we shall use the notations ${}^{s}c^* =: x \to xc^*c$ and $ab^* =: x \to ab^*x$ for $a,b,c,x \in E$.

10.4. PROPOSITION. *Given a* J^*-*triple* $(E,*)$, *the closed real Lie subalgebra* A *generated in the Lie algebra* $P(E)$ *of all polynomial vector fields by* $Q =: \{c - {}^{s}c^*;\ c \in E\}$ *admits the topological direct sum decomposition*

$$A = L \oplus Q$$

where L *is the closed real subalgebra of* $L(E)$ *generated by the family* $\{icc^*;\ c \in E\}$.

Proof: We have

$$A = \bigvee_{n=1}^{\infty} Q^n$$

where

$$Q^1 =: Q \quad \text{and} \quad Q^{n+1} =: \bigvee_{\substack{k+\ell=n \\ k,\ell \geq 1}} [Q^k, Q^\ell] \qquad \text{for} \quad n \geq 2$$

Here $\bigvee$ denotes the real linear hull operation, as usually. Since

$$(c-{}^{s}c^{*})_{y}^{(1}z = -({}^{s}c^{*})_{y}^{(1}z = -2yc^{*}z$$

and, by axiom J_2 we have

$$[a-{}^{s}a^{*},\ b-{}^{s}b^{*}]x = -2xa^{*}(b-xb^{*}x)+2xb^{*}(a-xa^{*}x) = 2(ab^{*}-ba^{*})x$$

for all $x\in E$, by putting $a =: \frac{i}{2}c$ and $b =: c$, we obtain

$$icc^{*}\in\{[a-{}^{s}a^{*},\ b-{}^{s}b^{*}];\quad a,b\in E\}\subset \mathcal{Q}^{2}$$

for $c\in E$. Therefore

$$\mathcal{L}+\mathcal{Q}\subset \mathcal{A}$$

The sum in the left-hand side is topologically direct, because clearly $\mathcal{Q}$ and $\mathcal{L}$ are closed subspaces in $\mathcal{P}(E)$. Moreover, by writing

$$\mathcal{L}_0 =: \bigvee\{icc^{*};\quad c\in E\}$$

we obtain

$$2ab^{*}-2ba^{*} = i(a+ib)(a+ib)^{*}-i(a-ib)(a-ib)^{*}\in\mathcal{L}_0$$

for all $a,b\in E$, whence

$$\mathcal{Q}^{2} = \mathcal{L}_0$$

Let us define $\mathcal{L}_0^n$ for $n\in\mathbb{N}$ in the same manner as $\mathcal{Q}^n$. We claim that

$$(10.1)\qquad \mathcal{Q}^{2k-1}\subset\mathcal{Q}\qquad\text{and}\qquad \mathcal{Q}^{2k}\subset\bigvee_{r=1}^{k}\mathcal{L}_0^{r}$$

for $k = 1,2,\ldots$ Indeed, we have already established (10.1) for $k = 1$. Now we proceed by induction on k. Assume we have

$$\mathcal{Q}^{2s-1}\subset\mathcal{Q}\qquad\text{and}\qquad \mathcal{Q}^{2s}\subset\bigvee_{r=1}^{s}\mathcal{L}_0^{r}$$

for all $s = 1,2,\ldots,,k$. Then

$$Q^{2k+1} = \bigvee_{\substack{m+n=2k+1 \\ m,n\geq 1}} [Q^m, Q^n] = \bigvee_{\substack{s+t=k \\ s\geq 1, t\geq 0}} [Q^{2s}, Q^{2t+1}] \subset \bigvee_{r\leq k} [L_0^r, Q]$$

and also

$$Q^{2k+2} = \Big(\bigvee_{s+t=k+1} [Q^{2s}, Q^{2t}]\Big) \vee \Big(\bigvee_{s+t=k} [Q^{2s+1}, Q^{2t+1}]\Big) \subset$$

$$\subset \Big(\bigvee_{s+t\leq k+1} [L_0^s, L_0^t]\Big) \vee [Q,Q] \subset \Big(\bigvee_{r=1}^{k+1} L_0^r\Big) \vee L_0 = \bigvee_{r=1}^{k+1} L_0^r$$

Thus, it suffices to show that

$$[L_0^r, Q] \subset Q \qquad , \qquad r = 1,2,\ldots$$

For r=1 this relation means

$$[iaa^*, b-{}^s b^*] \in Q \quad , \qquad a,b \in E$$

which is indeed true, since $(aa^*)_x^{(1} = aa^*$ and so, by axiom J_1 we have

$$[iaa^*, b-{}^s b^*]x = iaa^*b - iaa^*(xb^*x) + 2xb^*(iaa^*x) = iaa^*b - ix(aa^*b)^*x =$$

$$= i(aa^*b-{}^s(aa^*b)^*)x$$

If we have $[L_0^s, Q] \subset Q$ for s= 1,2,...,r then, from the definition of L_0^r and the Jacobi identity, we derive

$$[L_0^{r+1}, Q] = \bigvee_{s+t=r+1} [[L_0^s, L_0^t], Q] \subset$$

$$\subset \Big(\bigvee_{s+t=r+1} [[L_0^s, Q], L_0^t]\Big) \vee \Big(\bigvee_{s+t=r+1} [L_0^s, [L_0^t, Q]]\Big) \subset$$

$$\subset \bigvee_{s+t=t+1} \Big([Q, L_0^t] \vee [L_0^s, Q]\Big) \subset \vee Q = Q$$

This completes the proof. #

10.5. COROLLARY. *We have* $[L,Q] \subset Q$. *Namely,*

$$\ell c - {}^{s}(\ell c)^{*} = [\ell,\ c - {}^{s}c^{*}]$$

i.e.

$$x(\ell c)^{*}x = \ell(xc^{*}x) - 2xc^{*}(\ell x) \qquad x \in E$$

for all $\ell \in L$ *and* $c \in E$.

Proof: Let $\ell \in L$ and $c \in E$ be given. Then $\ell = \lim_{n\to\infty} \ell_n$ where $\ell_n \in \bigvee_{k=1}^{n} L_0^k$. Hence $[\ell_n,\ c - {}^{s}c^{*}] \in Q$ for all $n \in \mathbb{N}$ and

$$[\ell,\ c - {}^{s}c^{*}] = \lim_{n\to\infty} [\ell_n,\ c - {}^{s}c^{*}] \in Q$$

From the linearity of ℓ it follows that

$$[\ell,\ c - {}^{s}c^{*}]x = \ell c - \ell(xc^{*}x) + 2xc^{*}(\ell x)$$

Thus, the vector field $v =: \ell c - {}^{s}(\ell c)^{*} - [\ell,\ c - {}^{s}c^{*}]$ belongs to Q and vanishes at 0. Therefore $v = 0$.

#

10.6. EXERCISE. Prove that if $(E,*)$ is a $*$-triple such that $L \oplus Q$ (defined as in proposition 10.4) is a Lie algebra of vector fields, then $(E,*)$ is a J^*-triple.

§2.- Polarization in J^*-algebras.

Before proceeding to geometrical considerations, we investigate the basic algebraic consequences of the J^*-triple axioms.

10.7. LEMMA. *Let* E, F *be complex vector spaces and* A_1, A_2: $E\times E \to F$ *real bilinear maps. Then, if* $s_{A_1} = s_{A_2}$ (i.e., $A_1(x,x) = A_2(x,x)$ *for all* $x \in E$) *and* A_1, A_2 *are either symmetric and complex bilinear or sesquilinear maps, then we have* $A_1 = A_2$.

Proof: It suffices to note that if A: $E\times E \to F$ is symmetric complex bilinear, then

$$A(x,y) = \frac{1}{4} A(x+y,\ x+y) - \frac{1}{4} A(x-y,\ x-y)$$

and, if $A: E\times E \to F$ is sesquilinear, then

$$A(x,y) = \frac{1}{4} \sum_{k=0}^{3} i^k A(x+i^k y,\ x+i^k y).$$

#

<u>10.8. PROPOSITION</u>. *Axiom* J_1 *is equivalent to the following evaluation formula:*

(J) $$x_1(f_1 a^* f_2)^* x_2 = (af_1^* x_1) f_2^* x_2 - af_1^*(x_1 f_2^* x_2) + (af_1^* x_2) f_2^* x_1$$

for all $a, f_1, f_2, x_1, x_2 \in E$.

Proof: The implication (J) => (J_1) is trivial. Conversely, assume we have (J_1), i.e.

$$x(aa^*f)^*x = (aa^*x)f^*x + (aa^*x)f^*x - aa^*(xf^*x).$$

This means that, for fixed $a, f \in E$, the symmetric bilinear mappings

$$A_1(x_1,x_2) =: x_1(aa^*f)^*x_2$$

$$A_2(x_1,x_2) =: (aa^*x_1)f^*x_2 + (aa^*x_2)f^*x_1 - aa^*(x_1 f^* x_2)$$

coincide for $x_1 = x_2$, and hence everywhere by lemma 10.7. Therefore, for any fixed x_1, x_2, $f \in E$, the sesquilinear mappings

$$A_1'(a,g) =: x_1(ga^*f)^*x_1$$

$$A_2'(a,g) =: (a\ g^*x_1)f^*x_2 + (a\ g^*x_2)f^*x_1 - a\ g^*(x_1 f^* x_2)$$

coincide for $a = g$, and hence for arbitrary $a, g \in E$ by lemma 10.7

<u>10.9. COROLLARY</u>. *Axiom* (J_2) *is a consequence of* (J_1).

Proof: Axiom (J_1) implies (J); thus in particular

$$x(ax^*b)x = 2(xa^*x)b^*x - xa^*(xb^*x)$$

But we have $ax^*b = bx^*a$, i.e. the terms a,b can be interchanged in the right-hand side. Therefore

$$2(xa^*x)b^*x - xa^*(xb^*x) = 2(xb^*x)a^*x - xb^*(xa^*x)$$

and so

$$(xa^*x)b^*x = xa^*(xb^*x).$$

#

<u>10.10. PROPOSITION</u>. *Axiom* (J_1) *is also equivalent to*

(J'): $\quad [aa^*, bb^*] = a(ab^*b)^* - (bb^*a)a^*, \qquad a,b\in E$

Proof: Indeed, (J) implies

$$b(aa^*b)^*x = (aa^*b)b^*x + (aa^*x)b^*b - aa^*(bb^*x) = (aa^*b)b^*x + bb^*(aa^*x) - aa^*(bb^*x)$$

Thus

$$[bb^*, aa^*]x = b(aa^*b)^*x - (aa^*b)b^*x$$

proving (J').

If we suppose that (J') holds, the polarization argument of lemma 10.7 yields

$$[a_1a_2^*, b_1b_2^*] = a_1(a_2b_1^*b_2)^* - (b_1b_2^*a_1)a_2^*$$

whence

$$a_1(a_2b_1^*b_2)^*x = [a_1a_2^*,\ b_1b_2^*]x + (b_1b_2^*a_1)a_2^*x =$$

$$= a_1a_2^*(b_1b_2^*x) - b_1b_2^*(a_1a_2^*x) + (b_1b_2^*a_1)a_2^*x =$$

$$= (b_1b_2^*x)a_2^*a_1 + (b_1b_2^*a_1)a_2^*x - b_1b_2^*(a_1a_2^*x)$$

which is (J) for $a_1 = x_1$, $a_2 = f_1$, $b_1 = a$, $b_2 = f_2$ and $x = x_2$.

#

§3.- Flat subsystems.

10.11. DEFINITION. *Let* $(F,*)$ *be a real* J^**-triple. We say that a set* $S \subset F$ *is a "flat system" if* $ab^* = ba^*$ *for all* $a,b \in S$.
We say that S *is "commutative or abelian" if* $aa^* \smile bb^*$ *(i.e.* aa^* *commutes with* bb^**) for all* $a,b \in S$.
Finally, S *is "associative" if*

$$(x_1a^*x_2)b^*x_3 = x_1a^*(x_2b^*x_3)$$

for all $x_1, x_2, x_3 \in S$ and $a,b \in S$.
By passing to real linear combinations and using the polarization argument, it is easy to obtain the following

10.12. LEMMA. *Let* $(F,*)$ *be a real* J^**-triple and* $S \subset F$. *Then*

(a) S *is flat if and only if* $\vee S$ *is flat.*

(b) S *is abelian if and only if* $\vee S$ *is abelian, which occurs if and only if we have*

$$a_1b_1^* + b_1a_1^* \smile a_2b_2^* + b_2a_2^*$$

for all $a_1, b_1, a_2, b_2 \in S$.

(c) S *is associative if and only if* $\vee S$ *is associative.*

If (F,*) *is a complex* J^**-triple and* $S \subset F$ *then we also have.*

(b') S *is abelian if and only if* $^{\mathbb{C}}VS$ *is abelian, which occurs if and only if we have*

$$a_1 b_1^* \smile a_2 b_2^*$$

for all $a_1, b_1, a_2, b_2 \in S$.

(c') S *is associative if and only if* $^{\mathbb{C}}VS$ *is associative.*

Here $^{\mathbb{C}}VS$ denotes the complex linear hull of S.

10.13. EXERCISE. Prove the above lemma.

10.14. LEMMA. *Every complex flat* J^**-triple is abelian.*

Proof: Let $a, b \in F$; as F is flat we have

$$ab^* = ba^*$$

Since F is complex, $ia \in F$ whence

$$iab^* = b(ia)^* = -iba^* = -iab^*$$

and $ab^* = 0$. Then $a_1 b_1^* = a_2 b_2^* = 0$ for all $a_1, b_1, a_2\ b_2 \in F$ and F is abelian.

#

10.15. PROPOSITION. *Every real flat* J^**-triple is associative.*

Proof: Let $x_1, x_2, x_3 \in F$ and $a, b \in F$ be given. Applying axiom (J) and the fact that F is flat, we derive

$$(x_1 a^* x_2) b^* x_3 = b(x_1 a^* x_2)^* x_3 = (a x_1^* b) x_2^* x_3 + (a x_1^* x_3) x_2^* b - a x_1^* (b x_2^* x_3) =$$

$$= x_2 (a x_1^* b)^* x_3 + (x_1 a^* x_3) b^* x_2 - x_1 a^* (x_2 b^* x_3) =$$

$$= \left[(x_1 a^* x_2) b^* x_3 + (x_1 a^* x_3) b^* x_2 - x_1 a^* (x_2 b^* x_3) \right] +$$

$$+(x_1a^*x_3)b^*x_2-x_1a^*(x_2b^*x_3)$$

Substracting $(x_1a^*x_2)b^*x_3$ we easily derive

$$(x_1a^*x_3)b^*x_2 = x_1a^*(x_3b^*x_2)$$

#

§4.- Subtriples generated by an element.

10.16. DEFINITION. *Let* $(E,*)$ *be a* J^**-triple. A subtriple of* $(E,*)$ *is a closed subspace* F *of* E *such that* $FF^*F\subset F$, *i.e., we have* $xy^*z\in F$ *whenever* $x,y,z\in F$. *According as* F *is a real or a complex subspace of* E, *we speak of a real or a complex subtriple.*

Given any set $S\subset E$, *the smallest subtriple of* $(E,*)$ *containing* S *is called the* J^**-span of* S. *The smallest complex subtriple containing* S *is the* $^{\mathbb{C}}J^*$*-span of* S. *It is clear that*

$$J\text{-span } S = \bigvee_{n=0}^{\infty} S^n$$

where

$$S^0 =: \bigvee S \qquad \text{and} \qquad S^{n+1} =: \bigvee_{k+\ell+m=n} S^k(S^\ell)^* S^m$$

Similarly,

$$^{\mathbb{C}}J^*\text{-span } S = {}^{\mathbb{C}}\bigvee(J^*\text{-span } S)$$

Henceforth we assume that $(E,*)$ denotes a complex J^*-triple. Given $c\in E$, we shall write

$$F^c =: J^*\text{-span }\{c\} \quad , \quad E^c =: {}^{\mathbb{C}}J^*\text{-span }\{c\}$$

Furthermore, we set

$$c^n =: (cc^*)^{n-1}c$$

for n= 1,2,..., i.e., c^n is the n-th power of c with respect to c-multiplication.

<u>10.17. LEMMA</u>. *We have* $c^k\, c^{\ell *} = c^{k+\ell-1}c^*$ *for* $k,\ell = 1,2,..$

Proof: The stament is obvious for $k=\ell=1$. Suppose that, for some n, we have

$$c^k\, c^{\ell *} = c^r\, c^{s*} \tag{10.2}$$

whenever $k+\ell = r+s$ and $k, \ell, r, s \leq n$. Then, the system

$$S_n =: \{c^k;\ \ k= 1,2,\dots,n\}$$

is flat. Therefore

$$c^k\, c^{\ell *} \smile c^r\, c^{s*} \qquad k,\ \ell,\ r,\ s \leq n$$

Let $k, \ell \leq n$ and consider the term $c^k(c^{\ell+1})^*$. From axiom J we derive

$$c^k(c^{\ell+1})^*x = c^k(cc^*\, c^\ell)^*x =$$

$$= (cc^*\, c^k)c^{\ell *}x + (cc^*\, x)c^{\ell *}\, c^k - cc^*(c^k\, c^{\ell *}\, x) =$$

$$= c^{k+1}\, c^{\ell *}x - [c^k\, c^{\ell *},\ cc^*]x = c^{k+1}\, c^{\ell *}x$$

for all $x \in E$. Hence, for any $r \leq n$ we have

$$c^r(c^{n+1})^* = c^{r+1}\, c^{n*} = c^{r+2}(c^{n-1})^* = \dots = c^{n+1}\, c^{r*}$$

which is the relation (10.2) with n+1 instead of n.

#

<u>10.18. THEOREM.</u> F^c *is a flat subtriple of* (E,*) *and*

$$F^c = \bigvee_{n=1}^{\infty} c^n$$

Proof: By lemma 10.17, the system $S=:\{c^n;\ n\in\mathbb{N}\}$ is flat, and hence so is its closed linear span. It is clear that $\bigvee S\subset F^c$. On the other hand, from (10.2) we see

$$c^k\, c^{\ell *}\, c^m = c(c^{k+\ell-1})^* c^m = c^m (c^{k+\ell-1})^* c = c^{k+\ell+m-2}\, c^*\, c =$$
$$= c\, c^*\, c^{k+\ell+m-2} = c^{k+\ell+m-1}$$

for any k, ℓ, n= 1,2,... That is, $S\,S^*S\subset S$ whence $\bigvee S$ is a subtriple; thus $F^c\subset \bigvee S$.

#

10.19. COROLLARY. E^c *is an associative abelian subtriple and*

$$E^c = \bigvee_{n=1}^{\infty} c^n$$

§5.- JB*-triples and Hermitian operators.

10.20. DEFINITION. *We say that a* J^**-triple* $(E,*)$ *is "bounded", or that it is a* JB^**-triple, if there exists a bounded open neighbourhood* U *of the origin in* E *such that all vector fields* $c-{}^s c^*$ *for* $c\in E$ *are complete in* U.

Curiously, the letter B here refers to Banach's name and not to the adjective "bounded", because these structures are closely related to the Jordan-Banach algebras.

In the sequel we shall reserve the notations Q, L, L_0 and A to designate the families of vector fields (cf. Proposition 10.4)

$$Q=:\{c-{}^s c^*;\ \ c\in E\}\ ,\quad L_0=:\{icc^*;\ c\in E\}$$

$$L=:\text{Lie span}(L_0),\qquad A=:L\oplus Q$$

10.21. LEMMA. *If* $(E,*)$ *is a* JB^**-triple, then we can find a norm* $|\cdot|$ *on* E, *which is equivalent to the original one* $\|\cdot\|$, *such that, for every* $L\in L$, $\exp L$ *is an isometry for* $|\cdot|$.

Proof: Choose U in accordance with the above definition, i.e., $c-{}^{s^*}c_{|U} \in \mathrm{aut}U$ for all $c\in E$. Since autU is a Lie algebra of vector fields, we have $A_{|U}\in \mathrm{aut}U$ for all $A\in \mathcal{A}$. In particular, $L_{|U}\in \mathrm{aut}U$ for all $L\in \mathcal{L}$.

Let us consider the functional defined on E by

$$|v| =: \delta_U(0,v) \qquad v\in E$$

From lemma 3.9 we know that $|\cdot|$ is a continuous seminorm on E. Moreover, if $0<\delta$ and M are such that $\delta B(E) \subset U \subset MB(E)$, then

$$\frac{1}{M} \|v\| \leq |v| \leq \frac{1}{\delta} \|v\| \qquad v\in E$$

Given any $A\in\mathcal{A}$, we have $\exp(A_{|U})\in \mathrm{aut}U$ and therefore

$$\delta_U[0, (\exp A_{|U})_0^{(1}v] = \delta_U(0,v)$$

for all $v\in E$, i.e., $(\exp A_{|U})_0^{(1}$ is a $|\cdot|$-isometric linear operator. To complete the proof it suffices to remark that every $L\in\mathcal{L}$ is a linear operator, thus $(\exp L)_0^{(1} = \exp L$.

#

10.22. COROLLARY. *For all* $c\in E$ *and* $t\in\mathbb{R}$, $\exp(itcc^*)$ *is an isometry with respect to* $|\cdot|$.

10.23. EXERCISE. Prove lemma 10.21 elementarily by showing that the convex balanced hull $\mathrm{co}(\Delta U)$ of U is an exp-invariant balanced convex bounded negihbourhood of 0, and so its gauge can be taken as $|\cdot|$.

10.24. DEFINITION. *We say that a linear operator* $A\in\mathcal{L}(E)$ *is hermitian if* $\exp(itA)$ *is a* $\|\cdot\|$*-isometry for all* $t\in\mathbb{R}$. *We set* Her(E) *for the set of hermitian operators on* E.

Thus, $A\in \mathrm{Her}(E)$ if and only if $iA_{|B(E)}\in \mathrm{aut}B(E)$. This fact shows that Her(E) is a Lie subalgebra of $\mathcal{L}(E)$.

10.25. EXAMPLE. In a Hilbert space H we have

$$\mathrm{Her}(H) = \{A\in \mathcal{L}(H);\ \ A \text{ is selfadjoint}\} = \{A\in \mathcal{L}(H);\ (Ax|x)\in\mathbb{R}\quad \forall x\in H\}$$

where $<,>$ is the scalar product on H.

10.26. LEMMA. *A linear operator* $A\in \mathcal{L}(E)$ *is hermitian if and only if* $<\phi,Ax>\in\mathbb{R}$ *whenever* $x\in E$ *and* $\phi\in E'$ *are such that* $<\phi,x> = \|\phi\|\ \|x\|$.

Proof: Let us fix $x\in E$ and $A\in \mathcal{L}(E)$ arbitrarily, and consider the function $\varphi : \mathbb{R} \to \mathbb{R}$ given by

$$\varphi(t) =: \|x_t\| \qquad \text{where } x_t =: \exp(itA)x.$$

Since $t \to x_t$ is analytic and $y \to \|y\|$ is lipschitzian, φ is locally lipschitzian. Therefore $\varphi'(t)$ exists except for a subset $S(A,x)$ of $\mathbb{R}$ whose Lebesgue measure is null, and we have

$$\varphi(t_2) - \varphi(t_1) = \int_{t_1}^{t_2} \varphi'(t)\,dt \qquad t_1,t_2\in\mathbb{R}$$

Let us consider any $t\in\mathbb{R}\setminus S(A,x)$. We claim that, for all $\phi\in E'$, $\phi\neq 0$, with $<\phi,x_t> = \|\phi\|\ \|x_t\|$, we have

$$(10.3)\qquad \varphi'(t) = \mathrm{Re}< \frac{\phi}{\|\phi\|}\ ,\ i\,Ax_t>$$

Indeed, let ϕ be in the above conditions and put $\Psi =: \dfrac{\Psi}{\|\phi\|}$. For $s\in\mathbb{R}$ we have

$$\frac{1}{s}\,[\varphi(t+s) - \varphi(t)] = \frac{1}{s}\,(\|x_{t+s}\| - \|x_t\|) = \frac{1}{s}\,(<\Psi,x_{t+s}> - <\Psi,x_t>) =$$

$$= <\Psi,\ \frac{1}{s}\,(x_{t+s}-x_t)> = \mathrm{Re}<\Psi,\ \frac{1}{s}\,(x_{t+s}-x_t)>$$

Thus

$$\varphi'(t) = \lim_{s\to 0} \mathrm{Re}<\Psi,\ \frac{1}{s}\,(x_{t+s}-x_t)> = \mathrm{Re}<\Psi, iAx_t>$$

proving (10.3).

Now we prove the lemma. Let $A \in L(E)$ be such that

(10.4) $(y \in E,\ \phi \in E',\ <\phi,y> = \|\phi\|\ \|y\|) \Rightarrow <\phi,Ay> \in \mathbb{R}$

Fix any $x \in E$ and any $t \in \mathbb{R} \setminus S(A,x)$. By the Hanh-Banach theorem, there are functionals $\phi \in E'$, $\phi \neq 0$ such that $<\phi,x_t> = \|\phi\|\ \|x_t\|$. Now we can apply (10.3) to any of these functionals ϕ to obtain

$$\varphi'(t) = \mathrm{Re}< \frac{\phi}{\|\phi\|}, iAx_t>$$

As A is assumed to satisfy (10.4), we have $<\phi, Ax_t> \in \mathbb{R}$ whence $\varphi'(t) = 0$. Thus, φ' is null almost everywhere in $\mathbb{R}$ and φ is constant. Then

$$\|\exp(itA)x\| = \varphi(t) = \varphi(0) = \|x\|$$

Since $x \in E$ and $t \in \mathbb{R}$ were arbitrary, we get $A \in \mathrm{Her}(E)$.

Conversely, let $A \in \mathrm{Her}(E)$ be given and assume that $x \in E$ and $\phi \in E'$ are such that $<\phi,x> = \|\phi\|\ \|x\|$. Then we have $\varphi(t) = \|x_t\| = \|x\|$ for all $t \in \mathbb{R}$, so that φ is derivable everywhere in $\mathbb{R}$ and $\varphi' = 0$. Applying (10.3) to t=0 we get

$$0 = \varphi'(0) = \mathrm{Re}<\phi, i\,Ax>$$

whence $<\phi,Ax> \in \mathbb{R}$.

#

A fundamental fact concerning hermitian operators is the following result known as Sinclair's theorem. As usually, Sp(A) and $\rho(A)$ denote respectively the spectrum and the spectral radius of A.

10.27. THEOREM. *If* $A \in \mathrm{Her}(E)$ *then* $\mathrm{Sp}(A) \subset \mathbb{R}$ *and* $\rho(A) = \|A\|$.

Proof: Let A denote the closed complex subspace spanned in $L(E)$ by the family $\{I, A, A^2, \dots\}$. Then A is a commutative Banach subalgebra of $L(E)$. According to Gel'fand's theorem, there is a compact topological space K and a continuous algebra homomorphism $g: A \to C(K)$ such that

$$\text{ran } g(X) = \text{Sp}X$$

for all $X \in A$, where ran g(X) denotes the range of the function g(X). We have

$$\| g(X) \| = \max_{\xi \in \text{Sp}X} |\xi| = \rho(X) \leq \| X \|$$

i.e., g is a contraction. Thus

$$g(\exp itA) = g\left(\sum_{n=0}^{\infty} \frac{(it)^n}{n!} A^n \right) = \sum_{n=0}^{\infty} \frac{(it)^n}{n!} g(A)^n = \exp[itg(A)]$$

Let us fix $\xi \in \text{Sp}A$ arbitrarily. Then $\xi = g(A)z$ for some $z \in K$; therefore, as g is a contraction, we have

$$1 = \| \exp itA \| \geq \| g(\exp itA) \| = \| \exp itg(A) \| \geq |\exp[itg(A)]z| =$$

$$= |\exp(it\xi)| = \exp(-t\text{Im}\xi)$$

for all $t \in \mathbb{R}$. Thus $\text{Im}\xi = 0$, whence $\text{Sp}A \subset \mathbb{R}$.

In order to prove $\| \rho(A) \| = \| A \|$, we define

$$A_r =: \frac{r}{\rho(A)} A$$

for $r \in \mathbb{R}$ and observe that $A_r \in \text{Her}(E)$. It is easy to see that $\rho(A_r) = |r|$ and we recall that

$$\sin A_r =: \frac{1}{2i} [\exp(i A_r) - \exp(-i A_r)]$$

i.e.

$$\| \sin A_r \| \leq \frac{1}{2} + \frac{1}{2} = 1 \tag{10.5}$$

for all $r \in \mathbb{R}$. Since

$$\sin^{-1}\xi = \int_0^{\xi} \frac{d\xi}{\sqrt{1-\xi^2}} = \sum_{n=0}^{\infty} (-1)^n \binom{-1/2}{n} \int_0^{\xi} \xi^{2n} d\xi =$$

$$= \sum_{n=0}^{\infty} \frac{(-1)^n}{2n+1} \binom{-1/2}{n} \xi^{2n+1} \qquad \text{for } \xi \in [-1,1]$$

and since $(-1)^n \binom{-1/2}{n} > 0$ for all $n \in \mathbb{N}$, the theorem follows immediately if we establish that the series

$$\sum_{n=0}^{\infty} \frac{(-1)^n}{2n+1} \binom{-1/2}{n} (\sin A_r)^n$$

converges in $L(E)$ for all r, $0 \leq r < \pi/2$, and its sum satisfies

$$A_r = \sin^{-1}(\sin A_r) \qquad 0 \leq r < \pi/2 \tag{10.6}$$

Indeed, (10.5) and (10.6) imply

$$\| A_r \| \leq \sum_{n=0}^{\infty} \left| \frac{1}{2n+1} \left(\frac{-1/2}{n} \right) \right| \, \| \sin A_r \|^n \leq \sum_{n=0}^{\infty} \frac{(-1)^n}{2n+1} \binom{-1/2}{n} =$$

$$= \sin^{-1} 1 = \pi/2$$

for $0 \leq r < \pi/2$, whence

$$\| A_{\pi/2} \| = \lim_{r \uparrow \pi/2} \| A_r \| \leq \pi/2 = \rho(A_{\pi/2})$$

i.e., $\| A \| \leq \rho(A)$.

To prove (10.6), observe that

$$\lim_{n\to\infty} \left[\frac{1}{2n+1} \binom{-1/2}{n} \right]^{\frac{1}{2n+1}} = 1$$

Since, by Gel'fand's spectral radius theorem, $\rho(B) = \lim_{n\to\infty} \|B^n\|^{1/n}$ for $B\in L(E)$, the map

$$B \to \sin^{-1} B =: \sum_{n=0}^{\infty} \frac{(-1)^n}{2n+1} \binom{1/2}{n} B^{2n+1}$$

is well defined and holomorphic on the domain $\{B\in L(E);\ \rho(B)<1\}$. On the other hand, if $0<r<1<\pi/2$, then

$$Sp(\sin A_r) = \text{ran } g(\sin A_r) = \text{ran}[\sin g(A_r)] = \{\sin \xi;\ \xi\in \text{ran } g(A_r)\} =$$

$$= \{\sin \xi;\ \xi\in SpA\} = \{\sin \xi;\ \xi\in[-r,r]\} \subset [-\sin r, \sin r]$$

i.e., $\rho(\sin A_r)<1$. Thus, the mapping $r \to \sin^{-1}(\sin A_r)$ is analytic on the interval $(0,\pi/2)$.

For sufficiently small values of r, we have

$$\sin^{-1}(\sin A_r) = \sum_{n=0}^{\infty} \frac{(-1)^n}{2n+1} \binom{-1/2}{n} \left[\sum_{m=0}^{\infty} \frac{(-1)^m}{(2m+1)!} A_r^{2m+1} \right]^{2n+1} =$$

$$= \sum_{k=0}^{\infty} \alpha_k A_r^k$$

where

$$\alpha_k =: \sum_{n=0}^{\infty} \frac{(-1)^n}{2n+1} \binom{-1/2}{n} \sum_{\substack{(2m_1+1)+..+(2m_n+1)=k \\ m_1,\ldots,m_k\geq 0}} \prod_{j=1}^{n} \frac{(-1)^{m_j}}{(2m_j+1)!}$$

However, we have $\xi = \sin^{-1}(\sin \xi) = \sum_{k=0}^{\infty} \alpha_k \xi^k$ for $|\xi|<1$, whence $\alpha_1 = 1$ and $\alpha_k = 0$ for all $k\neq 1$, which proves (10.6).

§6.- Function model for E^c.

10.28. DEFINITION. *A* J^**-triple* $(E,*)$ *is said to be "Hermitian" if* $cc^*\in Her(E)$ *for all* $c\in E$.

In this terminology, lemma 10.21 can be rephrased as

10.29. LEMMA. *Every* JB^**-triple is topologically isomorphic to a hermitian one.*

Throughout this section, $(E,*)$ will be a fixed hermitian J^*-triple such that

$$\| xy^*z \| \leq M \| x \| \ \| y \| \ \| z \|$$

for all $x,y,z\in E$ and some $M\in\mathbb{R}$. We also fix an element $c\in E$, $c\neq 0$. We know already that $E = \overset{\mathbb{C}}{\bigvee_{n=1}^{\infty}} c^n$ is a commutative and associative complex subtriple of E. Therefore, by setting

$$x.y =: xc^*y \qquad (x,y\in E),$$

the structure $(E^c,\cdot)$ is a commutative Banach algebra, F^c is a closed real subalgebra of E^c and c^n is the n-th power of c in $(E^c,\cdot)$, so that the notation is not misleading. It is convenient to introduce the *multiplication operator representation* $R_c : E^c \to L(E^c)$ given by

$$R_c : x \to xc^*|_{E^c}$$

Obviously, R_c is a continuous algebra homomorphism.

10.30. LEMMA. *We have* $F^c(F^c)^*\subset Her(E)$; *in particular* $R_c x\in Her(E^c)$ *for all* $x\in F^c$.

Proof: It suffices to see that $c^m c^{n*}\in Her(E)$ for $m,n = 1,2,..$ But we have

$$4ic^m c^{n*} = [(ic^m)-^s(ic^m)^*, c^n-^s c^{n*}]\in[Q,Q]\subset L = \text{Lie span}\{iaa^*;\ a\in E\}\subset$$

$$\subset iHer(E)$$

since iHer(E) is a Lie subalgebra of $L(E)$.

#

In view of Sinclair's theorem, we have the following more precise picture of Gel'fand's representation for $(E^c, \cdot)$: We set

$$K =: Sp(cc^*|_{E^c}) \quad \text{and} \quad C_0(K) =: \{f \in C(K);\ f(0) = 0\}$$

(thus $C_0(K) \equiv C(K)$ if $0 \notin K$) with

$$\|f\|_K =: \max_{\eta \in K} |f(\eta)|, \quad f \in C_0(K)$$

As usually, we write id_K for the identity function on K.

10.31. THEOREM. *There is a unique Banach algebra homomorphism* G: $E^c \to C_0(K)$ *such that*

$$g(c) = id_K, \quad \operatorname{ran} g \subset \{\sqrt{|id_K|}f;\ f \in C_0(K)\}$$

and

$$(10.7) \qquad \|g(a)\|_K = \|cc^* aa^*|_{E^c}\|^{1/2}, \qquad a \in E^c$$

$$(10.8) \qquad \|aa^*|_{E^c}\| = \left\|\frac{1}{id_K} |g(a)|^2\right\|, \qquad a \in E^c$$

Proof: Let $A_0 \subset \mathbb{C}[t]$ be the algebra of the polynomials vanishing at 0. For $p(t) = \sum_{k=1}^{N} r_k t^k \in A_0$ let us define

$$p(c) =: \sum_{k=1}^{N} \gamma_k c^k$$

Thus, if $^-$ denotes the closure in $C_0(K)$, we have

$$E^c = \{p(c);\ p\in A_0\}^-$$

If $q(t)\in A_0$ has real coefficients, then $R_c q(c)\in Her(E^c)$ and, by Sinclair's theorem, we have

$$(10.9)\qquad \| R_c q(c) \| = \rho[R_c q(c)] = \rho[q(R_c(c))] = \max_{\eta\in SpR_c(c)} |q(\eta)| = \|q\|_K$$

If $p(t)\in A_0$ is arbitrary, then

$$p(c)p(c)^*|_{E^c} = (\sum_{k=1}^{N} \gamma_k c^k)(\sum_{\ell=1}^{N} \gamma_\ell c^\ell)^*|_{E^c} =$$

$$= \sum_{n=2}^{2N} \sum_{k+\ell=n} \gamma_k \bar{\gamma}_\ell c^k c^{\ell *}|_{E^c} =$$

$$= \sum_{n=2}^{2N} \sum_{k+\ell=n} \gamma_k \bar{\gamma}_\ell c^{n-1} c^*|_{E^c} = R_c[p_0(c)]$$

where $p_0(t) =: \frac{1}{t} |p(t)|^2$. From (10.9) we obtain

$$(10.8')\qquad \| p(c)\ p(c)^*|_{E^c} \| = \| \frac{1}{id_k} |p|^2 \|_K \quad ,\quad p\in A_0$$

Moreover, we have $|p|^2(t) = |p(t)|^2 = tp_0(t)$, i.e.

$$R_c(|p|^2(c)) = |p|^2(R_c(c)) = R_c(c)R_c(p_0(c))$$

whence, also by (10.9)

$$(10.7')\qquad \| p\|_K = \| \ |p|^2 \|^{\frac{1}{2}} = \| cc^* p(c)p(c)^*|_{E^c} \|^{\frac{1}{2}} \quad ,\quad p\in A_0$$

From (10.7') it follows that the mapping $g_0: A_0(c) \to \mathcal{C}_0(K)$ given by

$$g_0[p(c)] =: p \, , \qquad p \in A_0$$

is a well defined continuous algebra homomorphism and

$$(10.7'') \qquad \| g_0(a) \|_K = \| cc^* \, aa^*_{|E^c} \|^{\frac{1}{2}} \qquad a \in A_0(c)$$

$$(10.8'') \qquad \| aa^*_{|E^c} \| = \| \frac{1}{id_K} |g_0(a)|^2 \|_K \qquad a \in A_0(c)$$

It is an immediate consequence of (10.7") that g_0 admits a continuous extension to $E^c = A_0(c)^-$ satisfying (10.7). Now let $(p_n)_{n \in \mathbb{N}} \subset A_0$ be a sequence such that $p_n(c) \to a$. Then, by (10.9) we have

$$\| \frac{d}{id_K} |p_n|^2 - \frac{1}{id_K} |p_m|^2 \|_K = \| R_c[\frac{1}{t} |p_n(t)|^2 - \frac{1}{t} |p_m(t)|^2] \| =$$

$$= \| [p_n(c)p_n(c)^* - p_m(c)p_m(c)^*]_{|E^c} \| \to$$

$$\to \| (aa^* - aa^*)_{|E^c} \| = 0$$

Thus, the sequence $(\frac{1}{id_K} \, p_n)_{n \in \mathbb{N}}$ converges uniformly to some element of $C_0(K)$. But we have $p_n = g_0[p_n(c)] \to g(a)$ in $C_0(K)$, whence

$$\frac{1}{id_K} |g(a)|^2 = \lim_{n \to \infty} \frac{1}{id_K} |p_n|^2 \in C_0(K)$$

and (10.8) holds.

#

<u>10.32. COROLLARY.</u> *We have* $g(a_1 a_2^* a_3) = \frac{1}{id_K} g(a_1)\overline{g(a_2)}g(a_3)$ *for all* $a_1, a_2, a_3 \in E^c$.

Proof: Indeed, $c^k c^{\ell *} c^m = c^{k+\ell-1} c^* c^m =$
$= c^{k+\ell-1} . c^m = c^{k+\ell+m-1}$, i.e.

$$g[(\alpha c^k)(\beta c^\ell)^*(\gamma c^m)] = g(\alpha\bar{\beta}\gamma c^{k+\ell+m-1}) = \alpha\bar{\beta}\gamma \, id_k^{k+\ell+m-1} =$$

$$= \frac{1}{id_K} (\alpha \, id_K^k)(\overline{\beta \, id_K^\ell})(\gamma \, id_K^m) =$$

$$= \frac{1}{id_K} g(\alpha \, c^k)\overline{g(\beta \, c^\ell)} g(\gamma \, c^m)$$

for all $k,\ell,m\in\mathbb{N}$ $\alpha,\beta,\gamma\in\mathbb{C}$.

#

<u>10.33. EXERCISE</u>. Let $\xi\in K$, $\xi>0$ and $a, x_0 \in E^c$ be given and define

$$x_t =: \exp[t(a-{}^s a^*)]x_0, \quad \gamma_t =: g(x_t)\xi, \quad \alpha =: g(a)\xi.$$

Show that

$$\gamma_t = \frac{\sqrt{\xi} \dfrac{\alpha}{|\alpha|} \tanh \dfrac{t|\alpha|}{\sqrt{\xi}}}{1+\gamma_0 \dfrac{|\alpha|}{\alpha\sqrt{\xi}} \tanh \dfrac{t|\alpha|}{\sqrt{\xi}}}.$$

Hint: $\frac{d}{dt} \gamma_t = \alpha - \frac{1}{\xi} \bar{\alpha} \gamma_t^2$.

<u>10.34. DEFINITION</u>. *We set* $\|a\|_c =: \rho(aa^*_{|E^c})^{\frac{1}{2}}$ *for* $a\in E^c$ *and* $U^c =: \{a\in E^c;\ \|a\|_c < 1\}$. *Furthermore, for* $\alpha\in\mathbb{C}$ *and* $\xi\in\mathbb{R}\setminus\{0\}$, *we define the function* $q_{\alpha,\xi}: \mathbb{C}\to\mathbb{C}$ *by means of*

$$q_{\alpha,\xi}: \zeta \to \alpha - \frac{\bar{\alpha}}{\xi}\zeta^2$$

<u>10.35. LEMMA.</u> *Let* $\xi\in\mathbb{R}\setminus\{0\}$ *be given and suppose that* $D\subset\mathbb{C}$ *is a bounded open domain such that* $q_{\alpha,\xi}|_D \in \operatorname{aut} D$ *for all* $\alpha\in\mathbb{C}$. *Then we have* $\xi>0$ *and* $D=\sqrt{\xi}\,\Delta$.

Proof: Clearly

$$\operatorname{aut}D \supset \{[q_{\alpha,\xi}, q_{\beta,\xi}]_{|D};\ \alpha,\beta\in\mathbb{C}\}$$

Since

$$[q_{\alpha,\xi}, q_{\beta,\xi}]\zeta = \frac{2}{\xi}(\alpha\bar{\beta}-\beta\bar{\alpha})\zeta \qquad \zeta\in D$$

we get $\operatorname{aut}D \supset \{t\ell;\ t\in\mathbb{R}\}$ where ℓ is the vector field $\ell:\zeta\to i\zeta$. Thus, $\exp(t\ell)\zeta = e^{it}\zeta$ and $e^{it}D = D$ for $t\in\mathbb{R}$.

Let us consider any $\zeta_0\in\partial D$ and any $\alpha\in\mathbb{C}$, $\alpha\neq 0$. Then, we can find a sequence $(\zeta_n)_{n\in\mathbb{N}}\subset D$ such that $\zeta_n\to\zeta_0$. Write $\zeta_{n,t} =: \exp(tq_{\alpha,\xi})\zeta_n$ for $n\in\mathbb{N}$ and $t\in\mathbb{R}$. It is well know that, for some $\delta>0$, we have $\lim_{n\to\infty}\zeta_{n,t} = \zeta_{0,t}$ for $|t|<\delta$. Thus $\delta_{0,t}\in\bar{D}$ for $|t|<\delta$. On the other hand, $\zeta_{0,t}\notin D$ because $\zeta_{0,t}\in D$ would imply $\zeta_0 = \exp(-tq_{\alpha,\xi})\zeta_{0,t}\in D$ contradicting $\zeta_0\in\partial D$ and $D\cap\partial D = \phi$. It follows that $\zeta_{0,t}\in\partial D$ for $|t|<\delta$ and hence $|\zeta_{0,t}| = |\zeta_0|$. Indeed, if $|\zeta_{0,t}|\neq|\zeta_0|$, then we can choose $s\in(-\delta,\delta)$ such that

$$|\zeta_{0,s}| = \frac{1}{2}|\zeta_0| + \frac{1}{2}|\zeta_{0,t}|$$

Since $\zeta_{n,s}\to\zeta_{0,s}$, we have

$$\min\{|\zeta_0|,|\zeta_{0,t}|\}<|\zeta_{m,s}|<\max\{|\zeta_0|,|\zeta_{0,t}|\}$$

for some m. But then, there exists $u\in(-\delta,\delta)$ with $|\zeta_{0,u}|=|\zeta_{m,s}|$, i.e.

$$\zeta_{0,u}\in\{e^{i\theta}\zeta_{m,s};\ \theta\in\mathbb{R}\}\subset\bigcup_{\theta\in\mathbb{R}}e^{i\theta}D = D$$

contradicting $\zeta_{0,u}\in\partial D$.

If we had $\zeta_0 = 0$, then $\zeta_{0,t} = 0$, whence

$$0=\frac{d}{dt}(\zeta_0 t)=q_{\alpha,\xi}(\zeta_{0,t})=q_{\alpha,\xi}(0)=\alpha\neq 0$$

a contradiction. Thus $\zeta_0=0$. Now we have

$$0=\frac{d}{dt}\Big|_0|\zeta_{0,t}|^2=\frac{d}{dt}\Big|_0\zeta_{0,t}\bar{\zeta}_{0,t}=2\mathrm{Re}\,q_{\alpha,\xi}(\zeta_0)\bar{\zeta}_0=2\mathrm{Re}(\alpha\zeta_0-\frac{\bar{\alpha}}{\xi}\zeta_0^2\zeta_0)$$

Since we may replace ζ_0 by $e^{it}\zeta_0$ for any $t\in\mathbb{R}$, the argument of chapter 8, proposition 8.4, yields

$$\frac{\bar{\alpha}}{\xi}\zeta_0^2\bar{\zeta}_0=\bar{\alpha}\zeta_0$$

i.e. $\xi=|\xi_0|^2$ for all $\zeta_0\in\partial D$. This is possible only if $\xi>0$ and $D=\sqrt{\xi}\,D$ since D is bounded.

#

<u>10.36. THEOREM</u>. *Let us consider the following statements:*

(a) E^c *is a* JB^**-subtriple of* $(E,*)$.

(b) $Sp(cc^*|_{E^c})\subset\mathbb{R}_+$ *and* $aa^*a\neq 0$ *for all* $a\in E^c\setminus\{0\}$.

(c) $\|\cdot\|_c$ *is a norm on* E^c *and* $cc^*|_{E^c}$ *is injective.*

(d) U^c *is bounded.*

Then we have (a) => (b) => (c) *and* (a) <=> (b)+(d).

Proof: (a) => (b). Let U be a bounded neighbourhood of 0 in E^c such that $a-{}^sa^*|_U\in\mathrm{aut}U$ for all $a\in E^c$. Let us fix $\zeta\in K\setminus\{0\}$ arbitrarily and consider $D_\xi=:\{g(a)\xi;\ a\in U\}$. The mapping $a\to g(a)\xi$ is a continuous linear functional on E^c which is non identically null; consequently, D_ξ is a bounded open subset of $\mathbb{C}$. Given any $x_0\in U$ and $a\in E^c$, consider the orbit

$$x_t=:\exp t(a-{}^sa^*)x_0 \qquad t\in\mathbb{R}$$

We have $\frac{d}{dt}x_t=a-x_ta^*x_t$; therefore if we define

$$\zeta_t =: g(x_t)\xi \qquad \text{and} \qquad \alpha =: g(a)\xi,$$

we have

$$\frac{d}{dt}\zeta_t = q_{\alpha,\xi}(\zeta_t) \qquad t\in\mathbb{R}$$

Thus, $q_{\alpha,\xi}|_{D_\xi} \in \mathrm{atuD}_\xi$ whenever $\alpha\in\{g(a)\xi;\ a\in E^c\}$. Moreover,

$$\{g(a)\xi;\ a\in E^c\}\supset\{g(\gamma c)\xi;\ \gamma\in\mathbb{C}\} = \{\gamma\xi;\ \gamma\in\mathbb{C}\} = \mathbb{C}$$

As a consequence of the previous lemma, we have $\xi>0$ and $D_\xi = \sqrt{\xi}\Delta$.

On the other hand, the relation $aa^*a = 0$ entails

$$\frac{d}{dt}(ta) = a = a-(ta)a^*(ta) \qquad t\in\mathbb{R}$$

i.e.

$$\{\exp t(a-{}^s a^*)0;\ t\in\mathbb{R}\} = \mathbb{R}a\subset U$$

whence $a = 0$.

(b) => (c). We have

$$\rho(aa^*|_{E^c}) = \| aa^*|_{E^c}\| = \| \frac{1}{\mathrm{id}_K}|g(a)|^2\|_K \qquad a\in E^c$$

Thus, we have $\| a\|_c<1$ if and only if

$$\max_{\xi\in K\setminus\{0\}} \frac{1}{\mathrm{id}_K}|g(a)\xi|^2<1$$

whenever $|g(a)\xi|<\sqrt{\xi}$, i.e., $g(a)\xi\in\sqrt{\xi}\Delta$ for all $\xi\in K\setminus\{0\}$. That is,

$$U^c = \{a\in E^c;\ \| a\|_c<1\} = \bigcap_{\xi\in K\setminus\{0\}}\{a\in E\ ;\ g(a)\xi\in\sqrt{\xi}\Delta\}$$

As $\sqrt{\xi}\Delta$ is a convex set in $\mathbb{C}$, the linearity of g entails the convexity of U^c.

The functional $\|\cdot\|_c$ is obviously positive homogeneous, i.e. $\|\lambda a\|_c = |\lambda|\ \|a\|_c$. Furthermore, given $a \in E^c \setminus \{0\}$ we have $aa^*a \neq 0$ by hypothesis, and so

$$\|a_c\| = \rho(aa^*|_{E^c})^{\frac{1}{2}} = \|aa^*|_{E^c}\|^{\frac{1}{2}} > 0$$

Thus $\|\cdot\|_c$ is a norm on E^c.

Now, assume that $cc^*a = 0$, i.e., $R_c(c)a = 0$. Let us choose a sequence of polynomials $(p_n)_{n \in \mathbb{N}} \subset A_0$ such that $p_n(c) \to a$. By setting $q_n(t) =: \frac{1}{t}\ |p_n(t)|^2$, we have

$$p_n(c)p_n(c)^*a = q_n[R(c)]a = 0 \qquad n \in \mathbb{N}$$

whence, by taking the limit, we obtain $aa^*a = 0$ and $a = 0$.

(d)+(b) => (a). Indeed, U^c is an open neighbourhood of 0 in E^c since the map $a \to aa^*|_{E^c}$ is continuous and $U^c = \{a \in E^c;\ \|aa^*|_{E^c}\| < 1\}$. Let us show that $a - {}^s a^*|_{U^c} \in \mathrm{aut}\, U^c$ for all $a \in E^c$. Since polynomial vector fields are bounded on bounded sets, this means that the maximal solution of

$$(10.10) \qquad \frac{d}{dt}\, x_t = a - x_t a^* x_t \qquad x_0 = x$$

passes through the boundary of U^c for some $x \in U^c$. That is $x_{t_0} \in \partial U^c$ or, equivalently,

$$1 = \|x_{t_0}\|_c = \max \frac{1}{\sqrt{id_K}}\ |g(x_{t_0})|$$

for some $t_0 \in \mathbb{R}$, while

$$1 > \|x_0\|_c = \max \frac{1}{\sqrt{id_K}}\ |g(x_0)|$$

Thus we may fix $\xi\in K\setminus\{0\}$ such that

$$g(x_{t_0})\xi\in\partial(\sqrt{\xi\Delta}) \quad \text{and} \quad g(x_0)\xi\in\sqrt{\xi\Delta}$$

Let us write

$$\gamma_t =: g(x_t)\xi \text{ and } \quad \alpha =: g(a)\xi$$

as usually. From (10.10) it follows that

$$(10.10') \qquad \frac{d}{dt}\gamma_t = q_{\alpha,\xi}(\gamma_t) \qquad t\in\mathbb{R}$$

With the aid of chapter 8 proposition 8.4, it is easy to see that $q_{\alpha,\xi}$ is complete in $\sqrt{\xi\Delta}$. Since $\gamma_0\in\sqrt{\xi\Delta}$, it follows that $\gamma_{t_0}\in\sqrt{\xi\Delta}$, a contradiction.

(a) = (d). Let U be a bounded open neighbourhood of 0 in E^c such that $a-{}^s a^*|_U\in$ autU for all $a\in E^c$. We prove that $U^c\subset U$.

Given any $a\in E^c$, we have $\exp(a-{}^s a^*)0\in U$. Let us define

$$x_t =: \exp t(a-{}^s a^*)0 \qquad t\in\mathbb{R}$$

and, once $\xi\in K\setminus\{0\}$ has been fixed arbitrarily, set

$$\gamma_t =: g(x_t)\xi \quad \text{and} \quad \alpha =: g(a)\xi$$

Thus the functions $t\to x_t$ and $t\to\gamma_t$ satisfy the differential equations (10.10), (10.10') with the initial values $x = 0$ and $\gamma = 0$, respectively. Since we have already established the implication (a) => (b), we know that $\xi>0$. In this case, the explicit solution of (10.10') is given by

$$\gamma_t = \frac{|\alpha|}{\sqrt{\xi}}\tanh\frac{|\alpha|}{\sqrt{\xi}}\alpha \qquad t\in\mathbb{R}$$

That is, for $a \in E^c$ we have

$$(10.11) \qquad g[\exp(a-{}^s a^*)0] = \frac{g(a)}{\sqrt{id_K}} \tanh \frac{|g(a)|}{\sqrt{id_K}} g(a) ,$$

To conclude $U^c \subset U$ we must show that, for every $b \in U^c$, there exists some $a \in E^c$ such that $b = \exp(a-{}^s a^*)0$. Observe that, from the hypothesis (a), it follows that the Gel'fand representation g of E^c is injective. Indeed, (a) => (b) => (c) whence, in particular,

$$\| g(x) \| = \| cc^* xx^* |_{E^c} \| \geq \| cc^*(xx^* x) \| \neq 0$$

whenever $x \in E^c \setminus \{0\}$. Thus, fixing $b \in U^c$ arbitrarily, it suffices to find $a \in E^c$ such that

$$g(b) = \frac{|g(a)|}{\sqrt{id_K}} \tanh \frac{|g(a)|}{\sqrt{id_K}} g(a)$$

i.e.

$$(10.12) \qquad g(a) = \frac{|g(a)|}{\sqrt{id_K}} \tanh^{-1} \frac{|g(b)|}{\sqrt{id_K}} g(b)$$

We have

$$\tan^{-1} t = \sum_{n=0}^{\infty} \frac{1}{2n+1} t^{2n+1} \qquad |t|<1$$

Therefore

$$\frac{|g(b)|}{\sqrt{id_K}} \tanh^{-1} \frac{|g(b)|}{\sqrt{id_K}} g(b) = \sum_{n=0}^{\infty} \frac{1}{2n+1} \frac{|g(b)|^{2n}}{id_K^n} g(b) =$$

$$= \sum_{n=0}^{\infty} \frac{1}{2n+1} g[(bb^*)^n b]$$

since $\dfrac{|g(b)|}{\sqrt{\mathrm{id}_K}} \leq \|b\|_c < 1$ for $b \in U^c$. Moreover, the relation $b \in U^c$ implies also $\|bb^*|_{E^c}\| < 1$, whence we see that the mapping

$$\varphi(b) =: \sum_{n=0}^{\infty} \frac{1}{2n+1} (bb^*)^n b \quad , \quad b \in U^c$$

is well defined and we have

$$g[\varphi(b)] = \sum_{n=0}^{\infty} \frac{1}{2n+1} g[(bb^*)^n b]$$

Thus, the choice $a =: \varphi(b)$ suits (10.12). Consequently $U^c \subset U$.

In the course of the proof (a) => (b) we have seen that, for any $\xi \in K \setminus \{0\}$, we have $\xi > 0$ and

$$\{g(a)\xi;\ a \in U\} = D_\xi = \sqrt{\xi}\Delta$$

Hence, it readily follows that

$$\|a\|_c = \max_{\xi \in K \setminus \{0\}} \frac{1}{\sqrt{\xi}} |g(a)\xi| < 1, \qquad a \in U$$

i.e., $U \subset U^c$. Thus $U = U^c$, which completes the proof.

#

<u>10.37. COROLLARY</u>. *If* E^c *is a* JB^**-subtriple of* $(E,*)$, *then* $\|\cdot\|_c$ *is an equivalent norm on* E^c, *the unit ball of* $\|\cdot\|_c$ *is* U^c, *and* U^c *is the unique bounded open neighbourhood of* 0 *in* E^c *in which every vector field* $a - {}^s a^*$, $a \in E^c$, *is complete. Moreover,*

$$\varphi : b \to \varphi(b) =: \sum_{n=0}^{\infty} \frac{1}{2n+1} (bb^*)^n b, \quad b \in U^c$$

is a real bianalytic mapping of U^c *onto* E^c *whose inverse is given by*

$$\varphi^{-1}: a \to \varphi^{-1}(a) = \exp(a - {}^{s}a^{*})0, \quad a \in E^{c}$$

We close this section with the following *Jordan representation* of the JB^*-triples generated by a single element.

10.38. THEOREM. *If* E^c *is a* JB^**-subtriple of* E, *then the mapping* $j: a \to \frac{1}{\sqrt{id_K}} g(a)$ *is a topological* J^**-isomorphism of* $(E^c,*)$ *onto* $(C_0(K),*)$. *Here* $C_0(K)$ *is endowed with its natural triple product* $f_1 f_2^* f_3 =: f_1 \overline{f}_2 f_3$.

Proof: We have

$$j(a_1 a_2^* a_3) = \frac{1}{\sqrt{id_K}} g(a_1 a_2^* a_3) = \frac{1}{\sqrt{id_K}} \frac{1}{id_K} g(a_1)\overline{g(a_2)}g(a_3) =$$

$$= j(a_1)\overline{j(a_2)}j(a_3)$$

for all $a_1, a_2, a_3 \in E$. Thus j is a J^*-homomorphism. Furthermore,

$$\|j(a)\|_K = \max \frac{1}{\sqrt{id_K}} |g(a)| = \|a\|_c \qquad a \in E^c$$

Since the norm $\|\cdot\|$ is equivalent to $\|\cdot\|_c$ on E^c, it follows that the range of j is closed in $C_0(K)$ and j is a topological J^*-isomorphism of E^c onto $\mathrm{range}(j) \subset C_0(K)$. Observe that

$$j[p(c)] = \frac{1}{\sqrt{id_K}} p = \sqrt{id_K}\,\frac{p}{id_K}, \qquad p \in A_0$$

Thus, in view of the Weierstrass-Stone theorem, we have

$$\mathrm{range}(j) \supset \{j[p(c)];\ p \in A_0\}^- =$$

$$= \{\sqrt{id_K}\, q;\ q \in \mathbb{C}[t]\}^- \supset \{\sqrt{id_K}\, f;\ f \in C_0(K)\}^- \supset \{p;\ p \in A_0\}^- = C_0(K).$$

#

10.39. COROLLARY. *Every* JB^**-triple generated by a single element is topologically* J^**-isomorphic to* $(C_0(K),*)$ *for some compact subset* $K\subset\mathbb{R}_+$.

10.40. EXERCISES. (a) Let Ω be a compact space, $a\in\Omega$ and $C_a(\Omega)=:\{f\in C(\Omega);\ f(a)=0\}$. Prove that $(C_a(\Omega),*)$ is a JB^*-subtriple of $C(\Omega)$ with respect to the natural triple product.

(b) Assume that $\|\cdot\|_a$ is an equivalent norm on $C_a(\Omega)$. Prove that $(C_a(\Omega),\|\cdot\|_a,*)$ is a hermitian JB^*-triple if and only if $\|\cdot\|_a$ is a *lattice norm*, i.e.,

$|f\ |\leqslant|f_2|\ \Rightarrow\ \|f\|_a\leqslant\|f_2\|_a$.

§7.- $(E^c,*)$ as a commutative Jordan algebra.

Next we extend our considerations from the associative algebra $(E^c,*)$ to the non necessarily associative algebra $(E^c,*)$. As previously, $(E^c,*)$ is a fixed hermitian J^*-triple, $c\in E^c\setminus\{0\}$ is also fixed and we write $xy=:xc^*y$ $(x,y\in E)$ and $K=:Sp(cc^*)|_{E^c}$, respectively.

10.41. DEFINITION. *Let* A *be a commutative but non necessarily associative Banach algebra. We say that* A *is a commutative* J*-algebra if*

$$\text{(J.A)}\qquad u^2(uv)=u(u^2v)\qquad u,v\in A$$

where $u^2=:uu$.

Given a commutative J*-algebra* A, *the linear operator* $R\colon A\to L(A)$ *given by*

$$R(u)v=:uv\qquad u,v\in A$$

is the multiplication representation of A.

The importance of R derives from the fact that $L(A)$ is an associative algebra. The axiom (JA) can be stated as

(JA') $R(u^2) \smile R(u) \qquad u \in A.$

(i.e., $R(u^2)$ and $R(u)$ commute) in terms of R.

If A is not associative, in general we do not have $R(xy) = R(x)R(y)$. This difficulty can be overcome by sacrificing linearity. The *quadratric representation* $P: A \to L(A)$ given by

$$P(u) =: 2R(u)^2 - R(u^2) \qquad u \in A$$

is one of the most powerful tools in the study of J-algebras. It can be shown that $P(x)P(y) = P(xy)$ for $x, y \in A$; however, the usual proofs (see e.g. |6|) are not elementary.

10.42. PROPOSITION. *The algebra* $(E, \cdot)$, *i.e.*, E *with the* c-*multiplication, is a commutative* J-*algebra such that*

$$R(u^2)R(u) = R(u)R(u^2) = \frac{2}{3} R(u)^3 + \frac{1}{3} R(u)^3 \qquad u \in E \tag{10.13}$$

Proof: Obviously, $(E, .)$ is a commutative Banach algebra. To prove (10.13) we start from (J_2). We have

$$(cu*c)v*c = cu*(cv*c)$$

$$u[(cu*c)v*c]^*u = u[(cu*(cv*c)]^*u = u[(cv*c)u*c] = u[cv*(cu*c)]^*u$$

for any fixed $u, v \in E$ since the u and v-multiplications are commutative. Using axiom (J) we obtain

$$u[(cu*c)v*c]^*u = 2[v(cu*c)^*u] - v(cu*c)^*(uc*u) =$$

$$= 2[(uc*v)c*u + (uc*u)c*v - uc*(vc*u)]c*u -$$

$$- [(uc*v)c*(uc*u) + (uc*(uc*u))c*v - uc*(vc*(uc*u))] =$$

$$= 2[(uc*u)c*v]c*u - [(uc*u)c*(uc*v) + (uc*(uc*u))c*v -$$

$$-uc^*(uc^*u)c^*v)] = 2R(u)[R(u^2)v] - [R(u^2)(R(u)v) +$$

$$+R(u^3)v - R(u)(R(u^2)v] = \{3R(u)R(u^2) - R(u^2)R(u) - R(u^3)\}v$$

In a similar manner, we have

$$u[cu^*(cv^*c)]^*u = 2(uc^*u)(cv^*c)^*u - uc^*[u(cv^*c)^*u] =$$

$$= 2[(vc^*u^2)c^*u + (vc^*u)c^*u^2 - vc^*(uc^*u^2)] -$$

$$-uc^*[2(vc^*u)c^*u - vc^*u^2] = 2[R(u)(R(u^2)v +$$

$$+R(u^2)(R(u)v) - R(u^3)v] - R(u)[2R(u)(R(u)v) - R(u^2)v] =$$

$$= \{3R(u)R(u^2) + 2R(u^2)R(u) - 2R(u^3) - 2R(u)^3\}v$$

By a similar process, we obtain

$$u[(cv^*c)u^*c]^*u = 2(u(cv^*c)^*u)c^*u - u(cv^*c)^*u^2 =$$

$$= 2[2(vc^*u)c^*u - vc^*u^2]c^*u - [(vc^*u)c^*u^2 + c^*u^2 +$$

$$+(vc^*u^2)c^*u - vc^*u^3] = 2R(u)[2R(u)^2v - R(u^2)v] -$$

$$-[R(u^2)R(u)v + R(u)R(u^2)v - R(u^3)v] =$$

$$= \{4R(u)^3 - 3R(u)R(u^2) - R(u^2)R(u) + R(u^3)\}v$$

and, finally, again by using axiom (J):

$$u[cv^*(cu^*c)]^*u = 2(vc^*u)(cu^*c)^*u - vc^*[u(cu^*c)^*u] =$$

$$= 2[(uc^*(vc^*u))c^*u + u^2c^*(vc^*u) - uc^*((vc^*u)c^*u)] -$$

$$-vc^*[2(uc^*u)c^*u - uc^*(uc^*u)] = 2\{R(u)[R(u)(R(u)v)] +$$

$$+ R(u^2)(R(u)v) - (R(u)[R(u)(R(u)v)]\} - R(u^3)v =$$

$$= \{2R(u^2)R(u) - R(u^3)\}v$$

Thus, if we set

$A =: R(u^2)R(u)$ $B =: R(u)R(u^2)$

$C =: R(u)^3$ $D =: R(u^3)$

we have

$$3B-A-D = 3B+2A-2D-2C = 4C-3B-A+D = 2A-D$$

whence, by substracting the last term, we obtain

$$3B-3A = 3B-D-2C = 2D-3B-3A+4C = 0$$

Therefore, A= B and so

$$3A-D-2C = -2(3A-D-2C) = 0$$

and finally $A = B = \frac{2}{3}C = \frac{1}{3}D$.

#

<u>10.43. PROPOSITION</u>. *If* (A,·) *is a commutative* J-*algebra, then we have the following "polarization formula"*

$$R(u^2z) = 2R(uz)R(u) - 2R(u)R(z)R(u) + R(u^2)R(z)$$

for all $u, z \in A$.

Proof: Given any $v \in A$ and any $\tau \in \mathbb{R}$, by axiom (JA) we have

$$R[(u+\tau v)^2]R(u+\tau v) = R(u+\tau v)R[(u+\tau v)^2]$$

whence, by developing and comparing the coefficients of τ in both sides, we get

$$R(u^2)R(v)z + 2R(uv)R(u)z = 2R(u)R(uv)z + R(v)R(u^2)z$$

i.e.

$$u^2(vz) + 2(uv)(uz) = 2u[(uv)z] + v(u^2z)$$

This can be reformulated as

$$R(u^2)R(z)v+R(uz)R(u)v= \quad 2R(u)R(z)R(u)v+R(u^2z)v.$$

#

<u>10.44. COROLLARY.</u> *If* R(u) *and* R(z) *commute, then*

$$R(u^2z)= 2R(uz)R(u)+P(u)R(z).$$

<u>10.45. DEFINITION</u>. *Let* A *be a non necessarily associative Banach algebra. We define the "unital extension"* $\tilde{A}$ *of* A *by*

$$\tilde{A}=: \mathbb{C}\oplus A= \{\lambda\oplus a;\ \lambda\in\mathbb{C},\ a\in A\}$$

$$\|\lambda\oplus a\| =: |\lambda|+\|a\|$$

$$(\lambda\oplus a)(\mu\oplus b)=:\lambda\mu\oplus(\lambda b+\mu a+ab)$$

where $\lambda,\mu\in\mathbb{C}$ *and* $a,b\in A$.

As usually, we identify $0\oplus a$ with a, $a\in A$, and $\lambda\oplus 0$ with λ, $\lambda\in\mathbb{C}$, and we shall write $\lambda+a$ instead of $\lambda\oplus a$.

It is easy to check that $\tilde{A}$ is an algebra and 1 is the unique element $e\in\tilde{A}$ such that ex= x= xe for all $x\in\tilde{A}$.

<u>10.46. DEFINITION</u>. *We say that* $u\in A$ *is "invertible" if* u *admits a unique inverse with respect to the* , *multiplication. We shall denote the inverse of* u *by* u^{-1}

<u>10.47. LEMMA</u>. *The unital extension of a commutative* J-*algebra is a commutative* J-*algebra.*

Proof: Let A be a commutative J-algebra. It is immediate from the definition that $\tilde{A}$ is a commutative J-algebra. On the other hand, we have

$$\tilde{R}(\lambda+u)= \tilde{R}(\lambda)+\tilde{R}(u)= \lambda id_{\mathbb{C}}+\tilde{R}(u)$$

and $\tilde{R}[(\lambda+u)^2]= \tilde{R}(u^2+2\lambda u+\lambda^2)= \tilde{R}(u^2)+2\lambda\tilde{R}(u)+\lambda^2 id_{\mathbb{C}}$ for $\lambda\in\mathbb{C}$ and $u\in A$, where we have written $\tilde{R}$ to denote the multiplication representation on $\tilde{A}$. Thus it suffices to show that

$$R(u^2)\smile\tilde{R}(u) \qquad u\in A$$

If $\mu\in\mathbb{C}$ and $v\in A$, then

$$[\tilde{R}(u^2), \tilde{R}(u)](\mu+v) = [\tilde{R}(u^2), \tilde{R}(u)]\mu + [R(u^2), R(u)]v = \mu u^2 u - \mu u^2 u + 0 = 0$$

#

<u>10.48. DEFINITION</u>. *Throughout the remainder part of this section,* A *will denote a fixed unital commutative Banach algebra. We set*

$$u^n =: R(u)^n 1, \quad U =: \{x\in A;\ \| R(x) - \mathrm{id}\| < 1\}, \quad A^u =: \mathit{Span}\{u^n;\ n = 0,\ldots,1\}.$$

<u>10.49. LEMMA</u>. *We have* $u^{n+2} = uu^{n+1} = u^2 u^n$ *for* $n = 0,1,2,..$ *and* $u\in A$.

Proof: The case $n = 0$ is trivial. If we assume the statement to be valid for some $n\in\mathbb{N}$, then

$$u^{n+3} = R(u)^{n+3}1 = R(u)u^{n+2} = uu^{n+2} = R(u)R(u^2)u^n = R(u^2)R(u)u^n =$$
$$R(u^2)u^{n+1} = u^2 u^{n+1}.$$

#

<u>10.50. PROPOSITION</u>. *For all* $n = 0,1,\ldots$ *and all* $u\in A$, $R(u^n)$ *admits a representation as a polynomial on* $R(u)$ *and* $R(u^2)$. *In particular,* $R(u^n) \smile R(u^m)$ *for* $m,n = 0,1,2,\ldots$

Proof: The cases $n = 1,2$, are trivial. If we assume the result to be valid for some $n\in\mathbb{N}$, from the polarization formula we obtain

$$R(u^{n+1}) = R(u^2u^{n-1}) = 2R(uu^{n-1})R(u) - P(u)R(u^{n-1}) = 2R(u^n)R(u) -$$
$$-P(u)R(u^{n-1}) = 2p_n[R(u), R(u^2)]R(u) - [2R(u)^2 -$$
$$-R(u)]p_{n-1}[R(u), R(u^2)]$$

for suitable polynomials p_{n-1}, p_n.

#

<u>10.51. PROPOSITION</u>. *We have* $u^{m+n} = u^m u^n$ *for all* $u\in A$ *and all* $n = 0,1,2,..$ *In particular,* $A^u =: \mathit{Span}\{u^n;\ n = 0,1,2,\ldots\}$ *is an associative subalgebra of* A.

Proof: The stament is trivial for $n+m\leqslant 3$. Assume it to be true for some pair n, m with $m+n = d$; then

$$u^n u^{m+1} = R(u^n)R(u)u^m = R(u)R(u^n)u^m = R(u)(u^n u^m) = R(u)u^d = u^{d+1}$$

#

<u>10.52. LEMMA</u>. *For* $u \in U$, u *is invertible and* $u^{-1} = R(u)^{-1}1$. *Thus, the mapping* $U \to A^u$ *given by* $u \to u^{-1} = \sum_{n=0}^{\infty} (u-1)^n$ *is holomorphic on* U.

Proof: For $u \in U$ we have $\|R(u)-id\| < 1$, so that $R(u)$ is invertible and $R(u)^{-1} = \sum_{n=0}^{\infty} [R(u)-id]^n$. Now, the relation $uv=1$ is equivalent to $R(u)v = 1$, i.e, $v = R(u)^{-1}1$.

#

For $x \in U$, we introduce the auxiliary function $H(x): U \to \mathcal{L}(A)$ given by $u \to -(u^{-1})_x^{(1}$.

<u>10.53. THEOREM</u>. *We have* $P(x) = R(x)R(x^{-1})^{-1} = H(x)^{-1}$ *for all* $x \in U$.

Proof: Let $x \in U$ and $y \in A$ be given. Then

$$-H(x)y = \frac{d}{dt}\Big|_0 (x+ty)^{-1} = \frac{d}{dt}\Big|_0 R(x+ty)^{-1}1 =$$

$$= \frac{d}{dt}\Big|_0 [R(x)+tR(y)]^{-1}1 = \frac{d}{dt}\Big|_0 [id+tR(x)^{-1}R(y)]^{-1}R(x)^{-1}1 =$$

$$= -R(x)^{-1}R(y)R(x)^{-1}1 = -R(x)^{-1}yx^{-1} = -R(x)^{-1}R(x^{-1})y = -R(x^{-1})R(x)^{-1}y$$

whence $H(x) = R(x^{-1})R(x)^{-1}$. Moreover, $H(x)$ commutes with any operator of the form $\sum_{-n \leq k \leq m} \alpha_k R(x)^k$ with $n, m \in \mathbb{N}$

Since x^{-1} and x^2 belong to A^x, we have

$$x^{-1}x^2 = x^{-1}(xx) = (x^{-1}x)x = x \qquad x \in U$$

Therefore, the functions $f: U \to A$ and $g: U \to A$ given respectively by $z \to z^{-1}$ and $z \to z^2$ satisfy

$$y = (fg)_x^{(1}y = (f_x^{(1}y) \cdot g(x) + f(x) \cdot (g_x^{(1}y)$$

that is,

$$y = [-H(x)y]x^2 + x^{-1}(2xy) \qquad y \in A$$

so that

$$\mathrm{id} = -R(x^2)H(x)+2R(x^{-1})R(x) = -R(x^2)H(x)+2R(x^2)H(x) = P(x)H(x) =$$
$$= H(x)P(x).$$

#

10.54. PROPOSITION . *We have* $[P(x)y]^{-1} = P(x)^{-1}y^{-1}$ *for all* $x,y \in U$.

Proof: Let $x \in U$ and $y \in A$ be given; then

$$x^{-1}P(x)y = R(x^{-1})R(x^{-1})^{-1}R(x)y = R(x)y = xy$$

Now, let us fix any $y \in U$ and define the functions $f: U \to U$ and $g: U \to A$ by means of $z \to z^{-1}$ and $z \to P(z)y$ respectively. Then $g_x^{(1}y^{-1}$ is given by

$$g_x^{(1}y^{-1}: z \to [2(z(zy))-z^2y]_x^{(1}y^{-1} = 2y^{-1}(xy)+2(x(y^{-1}y))-2(xy^{-1})y =$$
$$= 2R(y^{-1})R(y)x+2x-2R(y)R(y^{-1})x = 2x$$

so that

$$1 = (fg)_x^{(1}y^{-1} = g(x)f_x^{(1}y^{-1}+f(x)g_x^{(1}y^{-1} = [P(x)y]\ [-H(x)y^{-1}]+x^{-1}(2x) =$$
$$= -[P(x)y]\ [P(x)^{-1}y^{-1}]+2$$

#

10.55. THEOREM. *(Fundamental formula). For every* $x,y \in A$

$$P[P(x)y] = P(x)P(y)P(x)$$

In particular $P(x^2) = P[P(x)1] = P(x)P(1)P(x) = P(x)^2, \quad x \in A$

Proof: As the expressions on both sides of the fundamental formula are polynomials of x,y, it suffices to prove it for $x,y \in U$. Let $x,y \in U$ be given; from porposition 10.54 we get

$$[(P(x)z)_y^{-1}] = [P(x)^{-1}z^{-1}]_y^{(1} \qquad x,z \in U\ ,\ y \in A$$

whence, by the definition of $H(x)$, we obtain

$$-H[P(x)y]P(x) = P(x)^{-1}[-H(y)]$$

so that, by theorem 10.53

$$P[P(x)y]^{-1}P(x)= P(x)^{-1}P(y)^{-1}$$

whence the result follows.

#

10.56. THEOREM. *Let* $w\in A$ *be given and assume that* $u,v\in A^w$. *Then*

$$P(uv)= P(u)P(v)$$

Proof: By proposition 10.51, A^w is an associative subalgebra of A; thus, for $x,y\in A^w$ we have

$$P(x)y= 2x^2y-x^2y= x^2y$$

and $P(x)\smile P(y)$. Therefore,

$$P(x^2y)= P[P(x)y]= P(x)P(y)P(x)= P(x)^2P(y)= P(x^2)P(y)$$

for all $x,y\in A$. Moreover, we have $(x^2)_1^{(1}= 2id$ and $1^2= 1$; therefore, by the inverse function theorem, there exists a neighbourhood $V\subset A^w$ of 1 such that each $v\in V$ can be written in the form $v= x^2$ for some $x\in A^w$. Thus

$$P(vu)= P(v)P(u)$$

for all u,v in V. Since both sides of the last equality are polynomials on u,v, that relation holds for all pairs $u,v\in A^w$.

#

10.57. EXERCISES. (1) Let (E,*) be a J^*-triple and put $Q(a): x\mapsto ax^*a$ for $a,x\in E$. Show that $Q[Q(a)b]= Q(a)Q(b)Q(a)$.

(2) By using the representation Q, prove that the fundamental formula is valid in (E,c^*) Hint: $P(u)= Q(u)Q(c)$.

§8.- Positive J^*-triples and the convexity of homogeneous circular domains.

We begin with the following simple but crucial observation.

10.58. LEMMA. *Let* (E,*) *be a* JB*-*triple and assume that* U *is a bounded open neighbourhood of* 0 *in* E *such that* $a-{}^sa^*|_U\in\text{aut}U$ *for all* $a\in E$. *Then*

$$U= \{c\in E;\ \ \rho(cc^*)|_{E^c})<1\}$$

and we have $Sp(cc^*|_{E^c}) \subset \mathbb{R}_+$ *for all* $c \in E$.

Proof: Let $c \in E$ be given. It is easy to see that E^c is a JB^*-triple, too, and that $V^c =: U \cap E^c$ is a bounded open neighbourhood of 0 in E^c such that

$$(10.14) \qquad \exp \mathbb{R}(b - {}^s b^*) V^c = V^c \qquad b \in E^c$$

By corollary 10.37, the only bounded open neighbourhood of 0 in E^c satisfying (10.14) is the set

$$W^c = \{x \in E^c ; \rho(xx^*|_{E^c}) < 1\}$$

Thus, we have

$$(10.15) \qquad U \cap E^c = V^c = \{x \in E^c ; \rho(xx^*|_{E^c}) < 1\}$$

Now, let $c \in E$ be such that $c \in U$. Then $c \in U \cap E^c$ and by (10.15) we have $\rho(cc^*|_{E^c}) < 1$. Conversely, let $c \in E$ be such that $\rho(cc^*|_{E^c}) > 1$. By (10.15) we have $c \in U \cap E^c$ and therefore $c \in U$ so that

$$U = \{c \in E ; \rho(cc^*|_{E^c}) < 1\}$$

On the other hand, for $c \in E$, E^c is a JB^*-triple and from theorem 10.36 we obtain $Sp(cc^*|_{E^c}) \subset \mathbb{R}_+$.

#

<u>10.59. LEMMA</u>. *Assume that* $(E,*)$ *is a* JB^*-*triple and let* $c \in E$ *and* $\lambda \in \mathbb{C} \setminus \{0\}$ *be given. Then we have* $\lambda \notin Sp(cc^*)$ *if and only if* $\lambda - c$ *is an invertible ement of the commutative* J-*algebra* $(\tilde{E}^c, c^*)$.

Proof: Set $K =: Sp(cc^*|_{E^c})$ and consider the Jordan representation $j: E^c \to C_0(K)$ of $(E^c,*)$ (cf. theorem 10.38 and corollary 10.39). The map j is a topological J^*-isomorphism and we have

$$j(ac^*b) = j(a)\sqrt{id_K}\, j(b) \qquad a,b\in E^c$$

Writing . for the c-multiplication in $\tilde{E}$,

$$(\lambda-c)\cdot(\mu+a) = \lambda\mu+(\lambda a-\mu c-c.a) \qquad \mu\in\mathbb{C},\quad a\in E^c$$

Thus λ-c is invertible in $(\tilde{E}^c,\cdot)$ if and only if

$$\lambda a+\lambda^{-1}c-c.a= 0$$

for some $a\in E^c$, which by the isomorphism j is equivalent to

$$\lambda f+\lambda^{-1}\sqrt{id_K} - id_K f= 0$$

for some $f\in C_0(K)$, i.e.,

$$(\lambda^{-1}\sqrt{id_K}) \,/\, (\lambda-id_K)\in C_0(K)$$

which is equivalent to $\lambda\notin K$.

10.60. PROPOSITION. *If* (E,*) *a* JB*-*triple, then we have*

$$\rho(cc^*) = \rho(cc^*|_{E^c}) \text{ and } Sp(cc^*)\subset\mathbb{R}_+$$

for all $c\in E$.

Proof: Let us fix $c\in E\setminus\{0\}$ arbitrarily and consider the multiplication representation R on the commutative unital J-algebra $(\tilde{E},c^*)$. By corollary 10.19 we know that the subset $\{R(a);\ a\in\tilde{E}^c\}$ is a commutative family in $L(\tilde{E})$. Therefore, the closed algebra R spanned in $L(\tilde{E})$ by $\{R(a);\ a\in\tilde{E}^c\}$ is a commutative unital Banach subalgebra of $L(\tilde{E})$. Let Ω be its spectrum and

$$\tilde{g}:\ R\to C(\Omega)$$

the Gel'fand representation of R.

Now, let $\zeta \epsilon Sp(cc^*)$ be given. We have

$$R(c): \lambda+x \to \lambda c+cc^*x , \qquad \lambda \epsilon \mathbb{C},\ x \epsilon E$$

Thus

$$\zeta \epsilon Sp(cc^*) = Sp[R(C)_{|E^c}] \subset SpR(c)$$

i.e., the operator $\zeta id_{\tilde{E}} - R(c)$ is not invertible in $L(\tilde{E})$, hence, it is not invertible in R. Therefore $\zeta \epsilon$range $\tilde{g}(R(c))$, i.e., we can choose some $\omega \epsilon \Omega$ such that

$$\tilde{g}[R(c)]\omega = \zeta$$

Let us define

$$\eta =: \tilde{g}[P(c)]\omega$$

where $P(c) =: 2R(c)^2 - R(c^2)$, and let ζ_1, ζ_2 be the roots of the equation $z^2 - 2\zeta z + n = 0$, i.e.

$$\frac{1}{2}(\zeta_1+\zeta_2) = \zeta \qquad\qquad \zeta_1\zeta_2 = \eta$$

For $j = 1,2$, we have

$$P(c-\zeta_j) = P(c-\zeta_j,\ c-\zeta_j) = P(c) - 2P(\zeta_j, c) + P(\zeta_j^2) = P(c) - 2\zeta_j R(c) + \zeta_j^2 id_{\tilde{E}}$$

and therefore

$$\tilde{g}[P(c-\zeta_j)]\omega = \eta - 2\zeta_j\zeta + \zeta_j = 0 .$$

Consequently, $P(c-\zeta_j)$ has no inverse in R. It follows that $c-\zeta_j$ is not an invertible element of $\tilde{E}^c$. Indeed, if we had

$$(c-\zeta_j)\cdot(a_j+\lambda_j)=1$$

for some $a_j \in E^c$ and $\lambda_j \in \mathbb{C}$, then it would follow $P(a_j+\lambda_j)\in R$ and

$$P(c-\zeta_j)P(a_j+\lambda_j)=P(a_j+\lambda_j)P(c-\zeta_j)=P[(a_j+\lambda_j)\cdot(c-\zeta_j)]=P(1)=1$$

i.e., the invertibility of $P(c-\zeta_j)$ in R. Thus, by lemma 10.59 we have

$$\zeta_j \in \{0\} \cup Sp(cc^*|_{E^c}) \qquad j=1,2$$

whence

$$0 \leq \frac{1}{2}(\zeta_1+\zeta_2)=\zeta \leq \frac{1}{2}\rho(cc^*|_{E^c})+\frac{1}{2}\rho(cc^*|_{E^c})$$

since $Sp(cc^*|_{E^c}) \subset \mathbb{R}_+$.

#

<u>10.61. COROLLARY.</u> *We have*

$$Sp(cc^*) \subset \{\frac{1}{2}(\zeta_1+\zeta_2);\ \zeta_1,\ \zeta_2 \in \{0\} \cup Sp(cc^*|_{E^c})\}.$$

<u>10.62. DEFINITION.</u> *Let* B *be a Banach space and* $A \in L(E)$. *We say that* A *is a "hermitian positive operator" if* $A \in Her(E)$ *and* $SpA \subset \mathbb{R}_+$. *We write* $Her_+(E)$ *for the set of hermitian positive operators of* $L(E)$.

Let $(E,*)$ *be a* J^**-triple. We say that* E *is a "hermitian positive* J^**-triple "if we have* $cc^* \in Her_+(E)$ *for all* $c \in E$.

Finally, for $c \in E$ *we set*

$$\|c\|_\infty =: \rho(cc^*)^{\frac{1}{2}}$$

<u>10.63. LEMMA.</u> *Let* E *be a Banach space,* $A \in L(E)$ *and*

$$D=: \{(\phi,x)\in E'\times E\ ;\ \langle\phi,x\rangle = \|\phi\|\ \|x\|\}$$

We have $A \in Her_+(E)$ *if and only if* $\langle\phi,x\rangle \geq 0$ *for all* $(\phi,x)\in D$.

Proof: Assume that $A\in Her_+(E)$ and write $r=: \rho(A)$. Observe that $A-\frac{r}{2}\, id_E \in Her(E)$ and

$$Sp(A-\frac{r}{2}\, id_E)\subset[-\frac{r}{2}, \frac{r}{2}]$$

From Sinclair's theorem we obtain

$$\| A-\frac{r}{2}\, id_E\| \leq \frac{r}{2}$$

Thus, if $(\phi,x)\in D$, then

$$\langle\phi, Ax-\frac{r}{2}x\rangle \geq -\| A-\frac{r}{2}\, id_E\| \geq -\frac{r}{2} = \langle\phi, -\frac{r}{2}x\rangle$$

whence $\langle\phi,Ax\rangle \geq 0$.

Conversely, assume that $\langle\Psi,Ay\rangle \geq 0$ for all $(\Psi,y)\in D$. Let $x\in E$ be arbitrarily fixed and define

$$x_t =: \exp(-tA)x \qquad \varphi(t) =: \| x_t\|$$

for $t\in\mathbb{R}$. Now we may apply (10.4) to the operator iA. It follows that $\varphi'(t)\leq 0$ for almost all $t\in\mathbb{R}$ and so the function φ does not increase. Thus $\|x_t\| \leq \|x\|$ for $t\in\mathbb{R}$, whence by the arbitrariness of $x\in E$, we obtain

$$\| \exp(-tA)\| \leq 1 \qquad t\geq 0 \tag{10.16}$$

By lemma 10.26, A is hermitian; thus by Sinclair's theorem we have $Sp(A)\subset\mathbb{R}$. Therefore, if we had $\lambda\in(SpA)\setminus\mathbb{R}_+$ for some $\lambda\in\mathbb{C}$, then we would have $\lambda>0$. However, in this case, from (10.16) and the spectral mapping theorem we would obtain

$$1<\exp(-\lambda)\leqslant\rho\,[\exp(-A)]\leqslant\|\exp(-A)\|\leqslant 1$$

which is a contradiction. Thus $Sp(A)\subset\mathbb{R}_+$.

<u>10.64. COROLLARY.</u> *If* $A\in Her(E)$, *then*

$$\|A\| = \sup_{(\phi,x)\in D} |\langle\phi,Ax\rangle|$$

Proof: Let us write

$$r_1 =: \inf_{(\phi,x)\in D} \langle\phi,Ax\rangle \qquad\qquad r_2 =: \sup_{(\phi,x)\in D} \langle\phi,Ax\rangle$$

Then $A-r_1 id_E$ and $r_2 id_E - A$ are positive hermitian operators, whence $Sp(A)\subset[r_1,r_2]$. Therefore $\|A\| = \rho(A) = \max\{|r_1|,|r_2|\}$.

#

<u>10.65. EXERCISES</u>. (a) Show that $Her_+(E)$ is a closed convex cone in $L(E)$.

(b) $A\in Her_+(E) \iff (\forall\zeta\in\mathbb{C},\ \ Re\zeta< 0 \Rightarrow \|\exp(\zeta A)\|\leqslant 1)$.

<u>10.66. PROPOSITION</u>. *If* (E,*) *is a hermitian positive* J^*-*triple, then* $\|\cdot\|_\infty$ *is a continuous seminorm on* E.

Proof: We have $cc^*\in Her_+(E)$ for $c\in E$. Then, by the definition of $\|\cdot\|_\infty$ and Sinclair's theorem

$$\|c\|_\infty = \rho(cc^*)^{\frac{1}{2}} = \|cc^*\|^{\frac{1}{2}}$$

On the other hand, the functional $c\to\|cc^*\|^{\frac{1}{2}}$ is continuous. Let D be as in lemma 10.63 and write

$$\langle a,b\rangle_{\phi,x} =: \langle\phi,ab^*x\rangle$$

for $(\phi,x)\in D$ and $a,b\in E$. Then, by corollary 10.64 we have

$$\|c\|_\infty = \sup_{(\phi,x)\in D} \langle c,c\rangle^{\frac{1}{2}} \qquad c\in E$$

For fixed $(\phi,x)\in D$, the mapping $(a,b) \to \langle a,b\rangle_{\phi,x}$, $a,b\in E$, is a semiinner product on E because, due to lemma 10.63, we have

$$\langle c,c\rangle_{\phi,x} = \langle \phi, cc^*x\rangle \geq 0 \qquad c\in E$$

Thus, $c \to \langle c,c\rangle_{\phi,x}$ is a seminorm on E for each $(\phi,x)\in D$. The supremum of a family of seminorms is always a seminorm.

#

As a summary of the chapter we get the following result due to Kaup.

10.67 THEOREM. *Let* $(E,*)$ *be a* JB^**-triple. Then*

(a) $\|\cdot\|_\infty$ *is an equivalent norm on* E *and its unit ball* B_∞ *is the only bounded open neighbourhood of* 0 *in* E *in which all vector fields* $c-{}^s c^*$, $c\in E$, *are complete.*

(b) $(E,*)$ *is a hermitian positive* J^**-triple when endowed with the norm* $\|\cdot\|_\infty$ *and we have*

$$\|c\|_\infty = \|cc^*\|^{\frac{1}{2}} = \|cc^*c\|^{\frac{1}{3}} \qquad c\in E$$

(c) *The mapping* $c \to \exp(c-{}^s c^*)0$ *is a real bianalytic isomorphism of* E *onto* B_∞ *whose inverse is*

$$b \to \sum_{n=0}^{\infty} \frac{1}{2n+1} (bb^*)^n b.$$

Proof: As $(E,*)$ is a JB^*-triple, by proposition 10.60 we have

$$\rho(cc^*) = \rho(cc^*|_{E^c}) \qquad c\in E$$

Now, fix any $c\in E$, $c\neq 0$. Then, E^c is also a JB^*-triple, and, by theorem 10.36, the map

$$\| a\|_{\mathbb{C}} = \rho(aa^*|_{E^c})^{\frac{1}{2}} \qquad a\in E^c$$

is a norm on E^c. Thus, from the definition of $\|\cdot\|_\infty$ we get

$$\| c\|_\infty = \rho(cc^*)^{\frac{1}{2}} = \rho(cc^*|_{E^c}) = \| c\|_c \neq 0$$

so that $\|\cdot\|_\infty$ is a norm which, by proposition 10.66, is continuous on E. In particular,

$$B_\infty =: \{c\in E;\ \|c\|_\infty < 1\}$$

is an open neighbourhood of 0 in E. Now, as E is a JB^*-triple, there exists some bounded open neighbourhood U of 0 in E such that we have

$$\exp\mathbb{R}(c-{}^{s}c^*)U = U$$

for all $c\in E$. By lemma 10.58 such a neighbourhood must be

$$U = \{c\in E;\ \rho(cc^*|_{E^c}) < 1\}$$

i.e., U is the open unit ball of E with respect to the norm $\|.\|_\infty$. Therefore $\|\cdot\|_\infty$ and the norm $\|\cdot\|$ are equivalent. Thus, we have proved (a).

Now, as $U = B_\infty$ is a bounded domain, $\mathrm{aut}B_\infty$ is a Lie algebra. Since we have $c-{}^{s}c^*\in\mathrm{aut}B_\infty$ for $c\in E$, we have $cc^*\in\mathrm{aut}B_\infty$, too. Then, the operators cc^* are all hermitian and, by proposition 10.60, we have $\mathrm{Sp}(cc^*)\subset\mathbb{R}$, so that $(E,*)$ is a hermitian positive triple when endowed with the norm $\|\ \|_\infty$. Then, by proposition 10.66, we have

$$\| c\|_\infty = \| cc^*\|^{\frac{1}{2}} \qquad c\in E$$

Moreover, from (10.8) and corollary 10.32 we obtain

$$\| cc^*c \|_\infty = \rho[(cc^*c)(cc^*c)^*]^{\frac{1}{2}} = \| (cc^*c)(cc^*c)^*{}_{|E^c} \|^{\frac{1}{2}} =$$

$$= \| \frac{1}{id_K} |g(cc^*c)|^2 \|_K^{\frac{1}{2}} = \| (\frac{1}{id_K} |g(c)|^2)^3 \|_K^{\frac{1}{2}} =$$

$$= \| \frac{1}{id_K} |g(c)|^2 \|_K^{\frac{3}{2}} = \| cc^*{}_{|E^c} \|^{\frac{3}{2}} = \| c \|^3_\infty$$

for c∈E. Thus, we have prove (b).

Finally, (c) is an immediate consequence of corollary 10.37.

#

<u>10.68. COROLLARY.</u> *Let* E *be a complex Banach space and assume that* D *is a (simply connected) bounded symmetric domain of* E. *Then, the Harish-Chandra realization* $\hat{D}$ *of* D *is convex. In particular, there is an equivalent norm* $\| \cdot \|_\infty$ *on* E *such that* D *is biholomorphically equivalent to the unit ball* B_∞ *of* $(E, \| \cdot \|_\infty)$.

Proof: From chapter 9 we know that the Harish-Chandra realization $\hat{D}$ of D is a bounded balanced circular domain of E. Let us consider its associated J^*-triple (E,*) as given by example 10.2. Obviously (E,*) is a JB^*-triple and $\hat{D}$ is a bounded neighbourhood of the origin such that $c-{}^sc^*\in \mathrm{aut}\hat{D}$ for all c∈E. By theorem 10.67, $\hat{D}$ is the open unit ball of E with respect to the norm $\| \cdot \|_\infty$. In particular, $\hat{D}$ is convex and the Harish-Chandra realization gives a biholomorphic isomorphism of D onto $\hat{D}$.

#

<u>10.69. COROLLARY.</u> *Let* E *be a complex Banach space,* D⊂E *a bounded balanced circular domain and* $E_0 =: (\mathrm{aut}D)0$. *Then:*

(a) *The set* $(\mathrm{Aut}D)0 = D \cap E_0$ is convex.

(b) *If* $A_c \in \mathrm{aut}D$ *is the unique vector field with the properties* $(A_c)_0^{(0} = c$, $(A_c)_0^{(1} = 0$, *there is a real analytic isomorphism* $\phi: D\cap E_0 \to E_0$ *such that*

$$(\exp A_{\phi(a)})0 = a \qquad a\in D\cap E_0$$

Proof: From chapter 7 we know that $D\cap E_0$ is a bounded balanced circular domain. Let $(E\ ,*)$ be its associated J^*-triple. Then $D\cap E_0$ is a bounded neighbourhood of 0 in E_0 such that $(\exp A_c)(D\cap E_0) = D\cap E_0$ for all $c\in E$; thus $(E_0\ ,*)$ is a JB^*-triple and, by theorem 10.67, $D\cap E_0$ is the unit ball of E_0 for a suitable norm; in particular, $D\cap E_0$ is convex. Moreover, $c \to (\exp A_c)0$ is a real bianalytic isomorphism of E_0 onto $D\cap E_0$; thus, its inverse $\phi: D\cap E_0 \to E_0$ is a real analytic isomorphism which satisfies

$$(10.17) \qquad (\exp A_{\phi(a)})0 = a$$

for all $a\in D\cap E_0$.

#

§9.- Some properties of the topology of local uniform convergence.

Next we apply the previous results to the study of the topology $\mathcal{T}$ of local uniform convergence on the group AutD for a bounded balanced circular domain D.

10.70. PROPOSITION. *The mapping* T: AutD $\to$ AutD *given by*

$$f \to Tf =: f\circ \exp A_{\phi f^{-1}(0)} \qquad f\in \mathrm{Aut}D$$

takes its values in $\mathrm{Aut}^0 D$ *and is continuous at* id_D *when* AutD *is endowed with the topology* $\mathcal{T}$.

Proof: By (10.17) we have

$$(Tf)0 = (f\exp A_{\phi f^{-1}(0)})0 = ff^{-1}(0) = 0$$

so that $Tf\in Aut^0D$ for $f\in AutD$. Moreover, if we endow AutD with the topology $\mathcal{T}$, the mappings

$$\begin{array}{c} AutD \to D\cap E_0 \to E_0 \to autD \\ f \to f^{-1}(0) \to \phi f^{-1}(0) \to A_{\phi f^{-1}(0)} \end{array} \tag{10.18}$$

are known to be continuous. We know as well that, for a suitable neighbourhood $\mathcal{M}$ 0f 0 in autD, the mapping

$$\begin{array}{c} \mathcal{M}\subset autD \to AutD \\ A \to expA \end{array}$$

is continuous (actually, a homeomorphism) when AutD is equipped with the topology $\mathcal{T}_a$. In particular, since $\mathcal{T}_a \geq \mathcal{T}$,

$$\begin{array}{c} autD \to AutD \\ A_{\phi f^{-1}(0)} \to expA_{\phi f^{-1}(0)} \end{array} \tag{10.19}$$

is continuous for the topology $\mathcal{T}$ on autD. Therefore, the is composite T of (10.18) and (10.19) is continuous at id_D for $\mathcal{T}$.

#

<u>10.71. PROPOSITION.</u> *Let* $(c_n)_{n\in\mathbb{N}}$ *be a sequence in* E_0 *with* $c_n \to 0$. *Then the sequence*

$$f_n =: expA_{c_n}$$

converges to id_D *uniformly on* D.

Proof: For $x\in D\cap E_0$ and $n\in\mathbb{N}$, let $y(t;\, x,c_n)$ denote the maximal solution of the initial value problem

$$\frac{d}{dt}\, y(t) = c_n - Q_{c_n}[y(t), y(t)] \;,\; y(0) = x.$$

Then, we have $f_n(x) = y(1;x,c_n)$ for all $n \in \mathbb{N}$ and $x \in D \cap E_0$. Thus

$$(10.20) \qquad f_n(x) - x = \int_0^1 (c_n - Q_{c_n}[y(s,x,c_n), y(s,x,c_n)]ds$$

As $c \to Q_c$ is a continuous real linear mapping, we have

$$\| Q_c(y,y) \| \leq K \| c \| \ \| y \|^2$$

for some constant K and all $c \in E_0$, $y \in E_0$. As D is bounded and $y(s,x,c) \in D \cap E_0 \subset D$ for $s \in \mathbb{R}$, $x \in D \cap E_0$ and $c \in E_0$, from (10.20) we derive

$$\| f_n - id_D \|_D \leq \int_0^1 (1+KM^2) \| c_n \| ds = (1+KM^2) \| c_n \| \to 0$$

whence the result follows.

#

10.72. PROPOSITION. *Let* $(f_n)_{n \in \mathbb{N}}$ *be a sequence in* AutD *such that* $T \lim_{n\to\infty} f_n = id_D$. *Then we have* $\lim_{n\to\infty} f_n = id_D$ *uniformly on* D.

Proof: From $T \lim_{n\to\infty} f_n = id_D$ we obtain $f_n^{-1}(0) \to 0$ and therefore $\phi f_n^{-1}(0) \to 0$. Then, by proposition 10.71 we have

$$(10.21) \qquad g_n =: \exp A_{-\phi f_n^{-1}(0)} \to id_D \text{ uniformly on D.}$$

Moreover, from $T \lim_{n\to\infty} f_n = id_D$ and proposition 10.70 we derive

$$T \lim_{n\to\infty} h_n = T \lim_{n\to\infty} Tf_n = id_D$$

As the transformations h_n are linear, the latter entails

(10.22) $$h_n \to id_D \quad \text{uniformly on } D$$

On the other hand, from the definition h_n, g_n and T we get

$$h_n g_n = (Tf_n)\exp A_{-\phi f_n^{-1}(0)} = f_n \exp A_{-\phi f_n^{-1}(0)} \exp A_{\phi f_n^{-1}(0)} = f_n$$

Therefore, by (10.21) and (10.22) we have

$$\| f_n - id_D \|_D = \| h_n g_n - id_D \|_D \leqslant$$

$$\leqslant \| h_n g_n - g_n \|_D + \| g_n - id_D \|_D = \| h_n - id_D \|_{g_n(D)} +$$

$$+ \| g_n - id_D \|_D = \| h_n - id_D \|_D + \| g_n - id_D \|_D \to 0 .$$

#

<u>10.73. PROPOSITION.</u> *Let* D *be a bounded balanced circular domain* D *of a Banach space* E. *On the group* AutD, *the topology* $\mathcal{T}$ *coincides with the topology* $\mathcal{T}_u$ *of uniform convergence over* D.

Proof: Let $(f_n)_{n\in\mathbb{N}}$ be a sequence of AutD such that $\mathcal{T}\lim_{n\to\infty} f_n = f$. It suffices to show that $(f_n)_{n\in\mathbb{N}}$ converges to f uniformly on D. Since AutD is a topological group with respect to $\mathcal{T}$, we have $\mathcal{T}\lim_{n\to\infty} f^{-1}f_n = id_D$ and, by proposition 10.72, the sequence $(f^{-1}f_n)_{n\in\mathbb{N}}$ converges to id_D uniformly on D.

On the other hand, by theorem 7.39, for every $\varepsilon>0$ there exist $\eta>0$ such that f can be extended to a bounded holomorphic mapping $F: D_\varepsilon \to D_\eta$. If M is a bound for F on D_ε, from Cauchy's inequalities we obtain that F is $\frac{M}{\varepsilon}$ -lipschitzian on D. Then we have

$$\| f_n - f \|_D = \| ff^{-1}f_n - f \|_D \leqslant \frac{M}{\varepsilon} \| f^{-1}f_n - id \|_D \to 0 .$$

#

LIST OF REFERENCES AND SUPPLEMENTARY READING

|1| BONSALL, F.F., DUNCAN, J.: Complete normed algebras; Springer-Verlag, Berlin (1973).

|2| BOURBAKI, N.: Variétés differentielles et analytiques, Fascicule de résultats, Hermann, Paris (1967).

|3| BOURBAKI, N.: Groupes et algèbres de Lie, Hermann, Paris (1967).

|4| BRAUN, R., KAUP, W., UPMEIER, H.: On the automorphisms of circular and Reinhardt domains in complex Banach spaces; Manuscripta Math. 25 (1878) 97-133.

|5| BRAUN, R., KAUP, W., UPMEIER, H.: A holomorphic characterization of Jordan C^*-algebras; Math. Z. 161 (1978) 277-290.

|6| BRAUN, H. & KOECHER, M.: Jordan algebras, Springer-Verlag, Berlin (1966).

|7| CARATHÉODORY, C.: Über das Schwarzschen Lemma bei analytisches Funktionen von zwei komplexen Veränderlichen; Math. Ann. 97 (1826) 76-98.

|8| CARATHÉODORY, C.: Über die Geometrie der analytischen Abbildungen die durch analytische Funktionen von zwei Veränderlichen vermittelt werden; Abh. Math. Seminar Hamburg 6 (1928) 97-145.

|9| CARTAN, E.: Sur les domaines bornés de l'espace de n variables complexes; Abh. Math. Seminar Hamburg 11 (1935) 116-162.

|10| CARTAN, H.: Sur les fonctions des plusieurs variables complexes. L'iteration des transformations interieures d'un domaine borné; Math. Z. 35 (1932) 760-773.

|11| CARTAN, H.: Sur les groupes de transformations analytiques, Hermann, Paris (1935).

|12| DIEUDONNÉ, J.: Foundations of modern analysis, Pure and App. Math. 10, Academic Press, New York (1968).

|13| DUNFORD, N. and SCHWARTZ, J.T.: Linear Operators, Pure and Applied Math. VII, Interscience, New York (1958).

|14| FRANZONI, T.: The group of biholomorphic automorphisms in certain j^*-algebras; Annali di Mat. Pura ed App. 127 (1981) 51-66.

|15| FRANZONI, T., VESENTINI, E.: Holomorphic maps and invariant distances; North Holland, Amsterdam (1980).

|16| GREENFIELD, S., WALLACH, N.: Automorphisms groups of bounded domains in Banach spaces, Trans. Amer. Math. Soc. 166 (1972) 45-57.

|17| HARRIS, L.A.: Bounded symmetric domains in infinite dimensional spaces, Proceedings in infinite dimensional holomorphy, Lecture Notes in Math. 364, Springer-Verlag, Berlin (1973).

|18| HARRIS, L.A.: Analytic invariants and the Schwarz-Pick inequality; Israel Jour. of Math. 34 (1979) 177-197.

|19| HARRIS, L.A.: A generalization of C^*-algebras, Proc. London Math. Soc., 3rd series 4o (1981) 331-361.

|20| HARRIS, L.A.: Schwarz-Pick systems of pseudometrics for domains in normed linear spaces; Advances in holomorphy, North-Holland, Amsterdam (1979) 345-406.

|21| HARRIS, L.A., KAUP, W.: Linear algebraic groups in infinite dimensions, Illinois Jour. of Math. 21 (1977) 666-674.

|22| HARRIS, L.A.: Operator Siegel domain, Proced. Edimb. Math. Soc. Sect. A 79 (1977) 137-156.

|23| HAYDEN, T.L., SUFFRIDGE, T.J.: Biholomorphic maps in Hilbert spaces have fixed points, Pacific Jour. of Math. 38 (1971) 419-442.

|24| HERVÉS, J.: On linear isometries of Cartan's factors in infinite dimensions, preprint.

|25| HILLE, E., PHILLIPS, R.S.: Functional Analysis and semigroups, Colloq. Pub. vol. 31, Amer. Math. Soc., Providence R.I. (1957).

|26| HOFFMAN, K.: Banach spaces of analytic functions, Englewood Cliffs, N.J., Prentice Hall (1962).

|27| HOCHSCHILD, G.: The structure of Lie groups, Holden Day, San Francisco (1965).

|28| HOLMES, R.B.: Geometrical Functional Analysis and its Applications. Springer-Verlag, Berlin (1975).

|29| ISIDRO, J.M.: On the group of analytic automorphisms of the unit ball of J^*-algebras, preprint.

|30| ISIDRO, J.M., VIGUÉ, J.P.: The group of biholomorphic automorphisms of symmetric Siegel domains and its topology, Ann. Scu. Norm. Sup. di Pisa, to appear.

|31| KAUP, W.: Über das Randverhalten von holomorphen Automorphismen beschänter Gebiete, Manuscripta Math. 3 (1970) 257-270.

|32| KAUP, W.: Algebraic characterization of symmetric complex Banach manifold, Math. Ann. 228 (1979) 39-64.

|33| KAUP, W.: Über die Klassifikation der symmetrischen hermiteschen Mannigfaltigkeiten unendlicher Dimension, I, Math. Ann. 257 (1981) 463-486.

|34| KAUP, W.: Über die Klassifikation der symmetrischen hermiteschen Mannigfaltigkeiten undendlicher Dimension, II, Math. Ann. 262 (1983) 57-75.

|35| KAUP, W.: A Riemann mapping theorem for symmetric bounded domains in complex Banach spaces, Math. Z. 183 (1983) 503-529.

|36| KAUP, W.: Über die Automorphismen Grassmannscher Mannigfaltigkeiten unendilicher Dimension, Math. Z. 144 (1975) 75-96.

|37| KAUP, W.: On the automorphisms of certain symmetric complex manifolds of infinite dimensions, Anais Acad. Brasileira de Ciências 48 (1976) 153-163.

|38| KAUP, W.: Bounded symmetric domains in complex Hilbert spaces, Symposia Mathematica vol. 26, Academic Press, New York (1982) 11-21.

|39| KAUP, W., UPMEIER, H.: An infinitesimal version of Cartan's uniqueness theorem uniqueness theorem, Manuscripta Math. 22 (1977) 381-401.

|40| KAUP, W., UPMEIER, H.: Jordan algebras and symmetric Siegel domains in Banach spaces, Math. Z. 157 (1977) 179-200.

|41| KAUP, W., UPMEIER, H.: Banach spaces with biholomorphically equivalent unit balls are isomorphic, Proc. Amer. Math. Soc. 58 (1976) 129-133-

|42| KOECHER, M.: An elementary approach to bounded symmetric domains, Lecture Notes, Rice University, Houston (1979).

|43| LOOS, O.: Bounded symmetric domains and Jordan pairs, Lecture Notes, University of California at Irvine (1977).

|44| NACHBIN, L.: Topology on spaces of holomorphic mappings, Ergebnisse der Math. 47, Springer-Verlag, Berlin (1969).

|45| NARASIMHAN, R.: Several complex variables, Chicago, Lecture Notes in Mathematics, The University of Chicago Press, (1971).

|46| PIATECKII SHAPIRO, I.I.: Automorphic functions and the geometry of classical domains, Gordon Breach, New York (1969).

|47| PIATECKII SHAPIRO, I.I.: The geometry and classification of bounded homogeneous domains, Russ. Math. Surveys 20 (1968) 1-48.

|48| POTAPOV, V.P.: The multiplicative structure of J-contractive matrix functions, Amer. Math. Soc. Transl. 15 (1960) 131-243.

|49| STACHÓ, L.L.: A short proof that the biholomorphic automorphisms of the unit ball in certain L^p spaces are linear, Acta Sci. Math. 41 (1979) 381-383.

|50| STACHÓ, L.L.: A projection principle concerning biholomorphic automorphisms, Acta Sci. Math. 44 (1982) 99-124.

|51| STACHÓ, L.L.: On fixed points of holomorphic automorphisms, Annali di Mat. Pura ed App. 128 (1980) 207-225.

|52| STACHÓ, L.L.: Elementary operator-theoretical approach to the subgroups of U(n), preprint.

|53| SUNADA, T.: Holomorphic equivalence problem for bounded Reinhardt domains, Math. Ann. 235 (1978) 111-128.

|54| THULLEN, P.: Zu den Abbildungen durch analytichen Funktionen mehrerer komplexer Veränderlichen, Math. Ann. 104 (1831) 244-259.

|55| UPMEIER, H.: Über die Automorphismengruppen von Banach Mannigfaltigkeiten mit invarianter Metric, Math. Ann. 223 (1976) 279-288.

|56| VESENTINI, E.: Automorphisms of the unit ball, Proceedings of a Seminar on several complex variables, Cartona, Italy (1977).

|57| VESENTINI, E.: Variations on a theme of Carathédoory, Ann. Scuola Norm. Sup. Pisa 6 (1979) 39-68.

|58| VESENTINI, E.: Invariant distances in Banach algebras, Adv. in Math. 47 (1983) 50-73.

|59| VIGUÉ, J.P.: Le groupe des automorphismes analytiques d'un domaine borné d'un espace de Banach complexe. Application aux domaines bornés symmetriques; Ann. Scient. Ec. Norm. Sup. 4e série 9 (1976) 203-382.

|60| VIGUÉ, J.P.: Automorphismes analytiques des produits continus de domaines bornés; Ann. Scient. Ec. Norm. Sup. 4e série 11 (1978) 229-246.

|61| VIGUÉ, J.P.: Frontière des domaines bornés cerclés homogènes, C.R. Acad. Sci. Paris A 288 (1979) 657-660.

|62| VIGUÉ, J.P.:Les automorphismes analytiques isometriques d'une varieté complexe normée, Bull. Soc. Math. France 110 (1982) 49-73.

|63| VIGUÉ, J.P.: Sur la décomposition d'un domaine borné symmetrique en produit continu de domaines bornés symmetriques irreductibles, Ann. Scient. Ec. Norm. Sup. 4$^{\underline{e}}$ série 14 (1981) 453-463.

|64| VIGUÉ, J.P.: Les domaines bornés symmetriques d'un espace de Banach complexe et les systèmes triples de Jordan, Math. Ann. 229 (1977) 223-231.

|65| VIGUÉ, J.P.: La distance de Carathéodory n'est pas interieure, Result. der Math. 6 (1983) 100-104.

|66| VIGUÉ, J.P.: Sur les applications holomorphes isometriques pour la distance de Carathéodory, Ann. Scu. Norm. Sup. Pisa, Serie IV, 9 (1982) 255-261.

|67| VIGUÉ, J.P.: Domaines bornés symmetriques dans un espace de Banach complexe, Symposia Mathematica 26, Academic Press, New York (1982) 95-104.

|68| VIGUÉ, J.P.: Géodésiques complexes et points fixes d'applications holomorphes, Adv. in Math. 52 (1984) 241-247.

|69| VIGUÉ, J.P.: Automorphisms analytiques d'un domaine de Reinhardt borné d'un espace de Banach a base, Ann. Inst. Fourier, Grenoble, 34, 2 (1984) 67-87.

|70| VIGUÉ, J.P., ISIDRO, J.M.: Sur la topologie du groupe des automorphismes analytiques d'un domaine cerclé borné, Bull. Sc. Math. 2$^{\underline{e}}$ série 106 (1982) 417-426.

REFERENCES ADDED IN PROOF

|71| BARTON, T.: Biholomorphic equivalence of bounded Reinhardt domains, preprint.

|72| BARTON, T., DINEEN, S. & TIMONEY, R.: Bounded Reinhardt domains in Banach spaces, preprint.

|73| DINEEN, S., TIMONEY, R. & VIGUE, J.P.: Pseudodistances invariantes sur les domaines d'un espace localement convexe, preprint.

|74| ISIDRO, J.M.: Linear isometries of the spaces of spinors, preprint.

|75| LEMPERT, L.: La metrique de Kobayashi et la représentation des domaines sur la boule, Bull. Soc. Math. Fr. 109 (1981). 427-474.

|76| LEMPERT, L.: Holomorphic retracts and intrinsic metrics in convex domains, Analysis Mathematica 8 (1982) 257-261.

|77| ROYDEN, H. & WONG, P.: Carathéodory and Kobayashi metric on convex domains, preprint.

|78| VESENTINI, E.: Complex geodesics, Compositio Math. 44 (1981) 375-394.

|79| VESENTINI, E.: Invariant distances and invariant differential metrics in locally convex spaces, Banach center publications 8 (1982) 493-512.

|80| VESENTINI, E.: Complex geodesics and holomorphic maps, Symposia Math. 26 (1982) 211-230.

|81| VIGUÉ, J.P.: The Carathéodory distance does not define the topology, Proc. Amer. Math. Soc. 91 (1984) 223-224.

|82| VIGUÉ, J.P.: Sur la caractérisation des automorphismes analytiques d'un domaine borné, preprint.

|83| VIGUÉ, J.P.: Points fixes d'applications holomorphes dans un domaine borné convexe de $\mathbb{C}^n$, preprint.